SpringerBriefs in Physics

For further volumes:
http://www.springer.com/series/8902

Timothy J. Hollowood

Renormalization Group and Fixed Points

in Quantum Field Theory

Springer

Timothy J. Hollowood
Department of Physics
Swansea University
Swansea
UK

ISSN 2191-5423   ISSN 2191-5431 (electronic)
ISBN 978-3-642-36311-5   ISBN 978-3-642-36312-2 (eBook)
DOI 10.1007/978-3-642-36312-2
Springer Heidelberg New York Dordrecht London

Library of Congress Control Number: 2013932853

Printed on acid-free paper

Springer is part of Springer Science+Business Media (www.springer.com)

Preface

The purpose of this short monograph is to introduce a powerful way to think about quantum field theories. This conceptual framework is Wilson's version of the renormalization group. The only prerequisites are a basic understanding of QFTs along the lines of a standard introductory course: the Lagrangian formalism and path integral, propagators, Feynman rules, etc.

The discussion begins with the simplest theories of scalar fields and then tackles gauge theories. Finally, theories with supersymmetry are briefly considered because they are a wonderful arena for discussing the renormalization group as there are a few key properties that one can prove exactly. For this reason the last chapter will provide a very basic description of some of the features that SUSY theories have with regard to the renormalization group, although the discussion of SUSY itself will necessarily be very rudimentary.

I apologise in advance to those who have pioneered this subject as I have not attempted to make a comprehensive list of references. The references that are given are intended to point the reader to sources which do have comprehensive lists.

I would like to thank the organizers of BUSSTEPP, Jonathan Evans in Cambridge 2008 and Ian Jack in Liverpool 2009, for providing excellently run summer schools that enabled me to develop my idea to teach QFT with the renormalization group as the central pillar. I would also like to thank Aaron Hiscox, Dan Schofield, and Vlad Vaganov for careful readings of the manuscript and for making some useful suggestions.

Swansea, December 2012 Timothy J. Hollowood

Contents

Acronyms

IR	Infra Red
QFT	Quantum Field Theory
RG	Renormalization Group
SUSY	Supersymmetric/Supersymmetry
UV	Ultra Violet
VEV	Vacuum Expectation Value

Chapter 1
The Concept of the Renormalization Group

In the opening chapter we introduce the renormalization group (RG) and associated concepts in a general form in order that the complications of particular applications do not obscure the simplicity of the ideas. There are several forms of the RG but our approach is the one pioneered by Wilson and for which he won the Physics Nobel Prize in 1982 (Wilson 1983). This approach follows from a remarkably simple and intuitive idea and yields a very powerful way to think about quantum field theories (QFT).

1.1 Effective Theories

The key thread running through the RG is the way that phenomena on different distance, or equivalently energy/momentum, scales relate to one another. The basic idea is that if we want to describe phenomena on length scales down to μ^{-1},[1] then we should be able to use a set of variables defined at the length scale μ^{-1}. For example, hadrons and mesons are built from quarks. As long as we consider processes which occur at low enough energies then the description is best couched in terms of hadrons and mesons. However, when we start to consider processes at higher energies, i.e. shorter distances, then the quark degrees-of-freedom cannot be ignored. At low energies, the point is *not* that the quarks do not matter, but the only way that they do matter is to set various coupling constants and masses in the *effective theory* where the manifest degrees-of-freedom are the hadrons and mesons.

The notion of an *effective theory* will be fundamental to our discussion. The idea is that the description of the physical world on distance scales $> \mu^{-1}$ is most efficiently described by a theory where the degrees-of-freedom are defined around the scale μ^{-1}. In this case there are no unnecessary degrees-of-freedom and the description is in some sense optimal. The effective theory will usually break down in some way for length scales smaller than μ^{-1}. At these smaller scales, a new effective theory

---

[1] In the following μ is a momentum scale.

T. J. Hollowood, *Renormalization Group and Fixed Points*, SpringerBriefs
in Physics, DOI: 10.1007/978-3-642-36312-2_1, © The Author(s) 2013

is called for, containing new degrees-of-freedom. The important point is that the parameters of the first effective theory (coupling constants and masses, etc.) will be fixed by properties of the more basic underlying theory. So there will be a set of matching conditions at momentum μ between the two layers of description.

The basic underlying assumption here is the intuitive notion of a *separation of scales*. The important point is that to make a successful effective theory one must identify the physically relevant variables at the scale in which one is interested and then understand how these variables interact. This will involve various "couplings" which could in principle be calculated from first principles by using the underlying microscopic theory. As long as there are only a few effective variables and a few couplings, these details can be fixed experimentally and then one will have an effective theory that can be used to make predictions.

Now we make these ideas more specific. In the functional integral formalism, a QFT is defined by the action $S[\phi; g_i]$ which is a functional of the fields and an ordinary function of what may be an infinite number of parameters that we call coupling constants. These include all the mass terms as well as the strengths of all the interactions. The couplings $\{g_i\}$ can be thought of as a set of coordinates on *theory space*. A QFT is then defined in terms of a functional integral

$$\mathscr{Z} = \int [d\phi]\, e^{i S[\phi; g_i]}. \tag{1.1}$$

All the difficulties in defining a QFT are lurking in the definition of the measure on the fields $\int [d\phi]$. This is a very tricky thing to define because a classical field has an infinite number of degrees-of-freedom and it is by no means a trivial matter to integrate over such an infinity. In perturbation theory a symptom of the difficulties in defining the measure shows up in the divergences at high momenta that occur in loop integrals. These UV divergences occur when the momenta on internal lines become large and so are intimately bound up with the fact that the field has an infinite number of degrees-of-freedom and can fluctuate on all scales.

1.2 RG Flow

At least initially, in order to properly define the measure $\int [d\phi]$, we have to implement some cut-off procedure to tame the UV degrees-of-freedom of the theory, or equivalently, in perturbation theory to regulate the divergences that occur in loop diagrams. As we have said above, these UV or high energy divergences occur because the fields can fluctuate at arbitrarily small distances and in order to regulate the theory we have to somehow suppress these modes. Whatever way this is done inevitably introduces a new momentum scale μ, the *cut off*, into the theory. The genius of Wilson's approach to the RG is to turn this seemingly unattractive feature to an advantage.

Cut offs or Regulators

There are many ways of introducing a cut off, or regulator, into a QFT. For example, one can define the theory on a spatial lattice (after Wick rotation to Euclidean space). In this case μ^{-1} is the physical lattice spacing. Or one can suppress the high momentum modes, again in the Euclidean theory, by modifying the action or the measure. In these physical cut-off schemes, μ is momentum or energy scale in the Euclidean version of the theory. Therefore when we Wick rotate back to Minkowski space, μ is a space-like momentum scale. Another way to regularize in the context of perturbation theory involves taking the integrals over the Euclidean loop momenta that arise and analytically continuing in the space-time dimension; a procedure known as dimensional regularization. After the divergences have been isolated and removed the space-time dimension in loop integrals can be fixed to the physical value. Although this procedure is not very intuitive, it is relatively simple to implement and is the industry standard in particle physics.

Suppose we have some physical quantity $F(g_i; \ell)_\mu$ which depends on a length scale ℓ. In the spirit of an effective theory we must have $\ell > \mu^{-1}$. This quantity also depends on the cut-off scale μ. The theory of RG postulates that one can change the cut off of the theory in such a way that the physics on length scales $> \mu^{-1}$ remains constant. If this is to be true then inevitably the couplings in the action must change with μ. This idea can be summed up in the RG equation:[2]

1st RG eqn. $\quad F(g_i(\mu); \ell)_\mu = F(g_i(\mu'); \ell)_{\mu'}.$ $\qquad$ (1.2)

The functions $g_i(\mu)$ define the RG flow of the theory in the space of couplings. An example of a physical quantity is the physical mass of a state of the theory $m_{\text{phys}} = m_{\text{phys}}(g_i(\mu))_\mu$, in which case there is no dependence on a length scale ℓ. Notice that we must be careful to distinguish between a physical mass which is defined by the position of a pole in the propagator of the field and a mass term in the Lagrangian. The latter is subject to renormalization just like any other coupling as we shall describe.

The RG flow in (1.2) is conventionally thought of as being from the UV towards the IR, i.e. decreasing μ, but we shall often think about it in the other direction as well, towards the UV with μ increasing. In order that the RG equation (1.2) can hold, it is necessary that the space of couplings includes *all* possible couplings (necessarily an infinite number). The RG is non-trivial because in order to lower the cut off we somehow have to "integrate out" the degrees-of-freedom of the theory that lie

---

[2] If F is a correlation function then there will be additional wave-function renormalizations of the operator insertions required: see (1.12) below.

between energy scales μ and μ'. In general this is a difficult step, however, as we shall see, in a QFT we are in a very lucky situation due to the remarkable focusing properties of RG flows. The first RG equation (1.2) makes clear the rôle of the RG momentum scale μ. From what we said about effective theories, the optimal choice for the cut off μ would be precisely at the scale $\mu = \ell^{-1}$, the momentum scale of the phenomena one is interested in.

There is a second way to use the RG equation (1.2). We can perform the RG transformation to lower the cut off $\mu \to \mu/s$ for a parameter $s > 1$, and then re-scale the length of the whole theory by s^{-1} to return the cut off to μ. If we define the couplings g_i to be dimensionless, which is always possible by scaling with appropriate powers of μ, we have a second RG equation of the form

$$\textbf{2nd RG eqn.} \quad F(g_i(\mu); s\ell)_\mu = s^{-d_F} F(g_i(\mu/s); \ell)_\mu, \tag{1.3}$$

where d_F is the mass dimension of F. In this equation, we may actually take s to be either greater than or less than one. This second RG equation is very useful because it means that the IR (long distance) or UV (short distance) behaviours of a given theory with a fixed cut off are encoded in the RG transformation of the couplings towards the IR or backwards towards the UV, respectively. For example, if $s > 1$ then on the left-hand side we have the physical quantity at a scale $s\ell$, whereas on the right-hand side we have the same quantity at a scale ℓ but with the couplings running towards the IR $\mu \to \mu/s$. This second way of thinking of the RG is often the physically relevant one in particle physics when we want to compare the behaviour of a theory at different momentum scales. For instance, we often want to know how the observables of a theory behaves as we change the energy scale. Now we see that the RG via the second RG equation (1.3) is one way to formulate an answer.

Since the physical observables of the theory can be determined in principle from the action, the RG transformation itself follows from the way the action changes as the cut off changes. The action with a particular cut off is known as the Wilsonian Effective Action $S[\phi; \mu, g_i]$ and since it depends on the fields, as well as the couplings and the cut off, the RG transformation must be generalized to allow for a change in the normalization of the fields:

The Key RG Equation

$$S[Z(\mu)^{1/2}\phi; \mu, g_i(\mu)] = S[Z(\mu')^{1/2}\phi; \mu', g_i(\mu')], \tag{1.4}$$

where $Z(\mu)$ is known as the "wave-function renormalization" of the field. In the general case with many fields, $Z(\mu)$ is a matrix quantity that can mix all the fields. For the simplest kind of QFT involving a single scalar field the action can be written as the sum of a kinetic term and a linear combination of "operators" $\mathcal{O}_i(x)$ which are powers of the fields and their derivatives, *e.g.* ϕ^n, $\phi^n \partial_\mu \phi \partial^\mu \phi$, etc:[3]

$$S[\phi; \mu, g_i] = \int d^d x \left[-\frac{1}{2} \partial_\mu \phi \partial^\mu \phi + \sum_i \mu^{d-d_i} g_i \mathcal{O}_i(x) \right] \tag{1.5}$$

where d_i is the mass dimension of $\mathcal{O}_i(x)$. Notice that we have chosen the couplings to be dimensionless and therefore have had to insert the appropriate power of the cut off to make the dimensional analysis consistent. In the second way of thinking about the RG (1.3), the re-scaling is irrelevant because the cut off remains the same. The wave-function renormalization factor $Z(\mu)$ can be thought of as the coupling to the kinetic term since

$$S[Z^{1/2}\phi; \mu, g_i] = \int d^d x \left[-\frac{Z}{2} \partial_\mu \phi \partial^\mu \phi + \cdots \right], \tag{1.6}$$

although Z also appears in the other terms in the action as a multiplicative factor.

In practice one thinks about infinitesimal RG transformations which are encoded in the *beta functions* of the couplings defined as

$$\beta_i(g_j) = \mu \frac{dg_i(\mu)}{d\mu}. \tag{1.7}$$

It is important to realize that these beta functions are only functions of the couplings themselves and only depend on the cut off implicitly through the couplings. The running couplings then follow by integration of the beta-function equations above. The solution of the flow equations relates the couplings at two scales as

$$g_i(\mu) = g_i \left(g_j(\mu'), \mu/\mu' \right). \tag{1.8}$$

The fact that couplings always appear as combinations $\mu^{d-d_i} g_i$ in the action (1.5) means that the beta functions have the form

$$\mu \frac{dg_i}{d\mu} = (d_i - d)g_i + \beta_i^{\text{quant.}}(g_j). \tag{1.9}$$

The first term arises from the classical scaling implied by the powers of μ in the action, while the second piece comes from the non-trivial quantum part of the RG transformation. This part of the transformation involves a non-trivial integrating out of degrees-of-freedom in the functional integral as we shall see explicitly in Chap. 2. If $\hbar$ were re-introduced $\exp[iS] \to \exp[iS/\hbar]$, then the quantum piece would indeed vanish in the limit $\hbar \to 0$. We also define the *anomalous dimension* of a field ϕ in

---

[3] The use of the term "operator" or "composite operator" derives from a canonical quantization approach to QFT in which one builds a Hilbert space on which the fields becomes operator-valued. The language is still used in the functional integral approach when, strictly-speaking, the quantities are not operators.

terms of the quantity $Z(\mu)$ as

$$\gamma_\phi(g_i) = -\frac{\mu}{2}\frac{d\log Z(\mu)}{d\mu}. \tag{1.10}$$

The reason for the terminology is explained below (1.13).

In particle physics the ultimate physical observables are the S-matrix elements which determine the probabilities for each process. However, the S-matrix can be extracted from a more basic set of observables known as the Green's functions. These are correlation functions of fields evaluated by making insertions into the functional integral

$$\Gamma^{(n)}(x_1,\ldots,x_n;\mu,g_i(\mu)) = \int [d\phi]_\mu\, e^{iS[\phi;\mu,g_i(\mu)]}\,\phi(x_1)\cdots\phi_n(x_n). \tag{1.11}$$

The subscript μ on the measure here reminds us that the measure can depend on the cut off chosen. It follows from (1.4) that for these quantities that depend on the fields we must generalize (1.2) to take account of wave-function renormalization, yielding an RG equation of the form

$$Z(\mu)^{-n/2}\Gamma^{(n)}(x_1,\ldots,x_n;\mu,g_i(\mu)) = Z(\mu')^{-n/2}\Gamma^{(n)}(x_1,\ldots,x_n;\mu',g_i(\mu')). \tag{1.12}$$

We can also write a version of this equation in the spirit of the second RG equation (1.3) where the cut off remains fixed but physical lengths are scaled:

$$\Gamma^{(n)}(sx_1,\ldots,sx_n;\mu,g_i(\mu))$$
$$= \left[\frac{Z(\mu)}{Z(\mu/s)}s^{d-2}\right]^{n/2}\Gamma^{(n)}(x_1,\ldots,x_n;\mu,g_i(\mu/s)). \tag{1.13}$$

The factor $s^{n(d-2)/2}$ here, reflects the mass dimension of the n field insertions. This form of the RG equation allows us to see why γ_ϕ defined in (1.10) is known as the anomalous dimension. In order to see why, consider an infinitesimal transformation $s = 1+\delta s$, in which case the multiplicative factor in (1.13), for one of the insertions, can be written

$$\left[\frac{Z(\mu)}{Z(\mu/s)}s^{d-2}\right]^{1/2} = 1 + \left[\tfrac{1}{2}(d-2) + \gamma_\phi\right]\delta s. \tag{1.14}$$

The form here reflects the fact that an RG transformation is more than just a scaling transformation; however, the net effect is as if the physical scaling dimension of ϕ receives the anomalous contribution γ_ϕ. It is important to realize that, in general, γ_ϕ depends on the couplings and so will itself be an implicit function of the RG scale.

1.3 UV and IR Limits and Fixed Points

We have seen that the RG scale μ represents the momentum scale of physical interest. What is particularly important about RG flows $g_i(\mu)$ are their IR and UV limits; namely the two limits $\mu \to 0$ and $\mu \to \infty$, respectively, since according to (1.3) these tell us how the theory behaves at very long (IR) or short (UV) distances. In this way of thinking, all physical masses relative to the cut off, m/μ, increase as we flow towards the IR. If a theory has no massless particles (we say that the theory has a *mass gap*) then as μ decreases beyond the mass of the lightest particle there are effectively no physical degrees-of-freedom left to propagate at the low momentum scale μ. Hence, in the IR limit $\mu \to 0$ we have an empty theory with no propagating states. The other possibility is when the RG flow starts on the critical surface:

Critical Surface

The infinite dimensional subspace in the space-of-theories for which the mass gap vanishes. These theories consequently have a non-trivial IR limit in which only the massless degrees-of-freedom remain.

In this case, as $\mu \to 0$ the massless degrees-of-freedom will remain, and in most conventional theories the couplings flow to a fixed point of the RG $g_i(\mu) \to g_i^*$ where the beta functions vanish:[4]

$$\mu \frac{dg_i}{d\mu}\bigg|_{g_j^*} = 0. \tag{1.15}$$

Notice that the wave-function renormalization and anomalous dimensions do not need to vanish at a fixed point.

The theories at the fixed points are very special because as well as only having massless degrees-of-freedom they have no other couplings with a non-vanishing mass dimension. This means that they are actually scale invariant. However, as we shall argue, this scale invariance is naturally promoted to the group of conformal transformations and so the fixed point theories are also known as "conformal field theories" (CFTs). The (connected part of the) conformal group consists of Poincaré transformations along with scale transformations, known also as "dilatations", $x^\mu \to sx^\mu$, along with special conformal transformations

$$x^\mu \longrightarrow \frac{x^\mu + x^2 b^\mu}{1 + 2b \cdot x + x^2 b^2}, \tag{1.16}$$

---

[4] There are some exotic situations where the couplings flow to a limit cycle.

for a vector b^μ. The infinitesimal transformations for Lorentz, dilatations and special conformal transformations are

$$\delta x^\mu = \varepsilon^\mu{}_\nu x^\nu, \qquad \delta x^\mu = s x^\mu, \qquad \delta x^\mu = x^2 b^\mu - 2 x^\mu (x \cdot b), \qquad (1.17)$$

where $\varepsilon^{\mu\nu} = -\varepsilon^{\nu\mu}$, respectively. In any local QFT there exists an energy-momentum tensor $T_{\mu\nu}$ which appears in the Ward identity

$$\sum_{p=1}^{n} \langle \phi_1(x_1) \cdots \delta \phi_p(x_p) \cdots \phi_n(x_n) \rangle = - \int d^d x \, \langle \phi_1(x_1) \cdots \phi_n(x_n) T^\mu{}_\nu(x) \rangle \, \partial_\mu (\delta x^\nu) .$$

$$(1.18)$$

Invariance of the QFT—that is the vanishing of the left-hand side—under Lorentz transformations requires that the energy-momentum tensor is symmetric, $T_{\mu\nu} = T_{\nu\mu}$, while invariance under dilatations requires that it is traceless $T^\mu{}_\mu = 0$. These two conditions are then sufficient to imply invariance under infinitesimal special conformal transformations, since for the last variation in (1.17)

$$T^\mu{}_\nu \big(2 x_\mu b^\nu - 2 x^\nu b_\mu - 2(x \cdot b) \delta_\mu{}^\nu \big) = 2 T_{\mu\nu} \big(x^\mu b^\nu - x^\nu b^\mu \big) - 2(x \cdot b) T^\mu{}_\mu = 0.$$

$$(1.19)$$

Returning to the the RG flows, in the neighbourhood of a fixed point $g_i = g_i^* + \delta g_i$, we can always linearize the beta function:

$$\mu \frac{dg_i}{d\mu} \bigg|_{g_j^* + \delta g_j} = A_{ij} \delta g_j + \mathcal{O}(\delta g_j^2) . \qquad (1.20)$$

In a suitable basis for $\{\delta g_i\}$, which we denote by $\{\sigma_i\}$, the linear term is diagonal:

$$\mu \frac{d\sigma_i}{d\mu} = (\Delta_i - d)\sigma_i + \mathcal{O}(\sigma^2) \qquad (1.21)$$

and so to linear order the RG flow is simply

$$\sigma_i(\mu) = \left(\frac{\mu}{\mu'} \right)^{\Delta_i - d} \sigma_i(\mu') . \qquad (1.22)$$

The quantity Δ_i is called the scaling (or conformal) dimension of the operator associated to σ_i. In general, in an interacting QFT it will not be the mass dimension and the difference

$$\gamma_i = \Delta_i - d_i \qquad (1.23)$$

is the *anomalous dimension* of the operator to mirror the definition of the anomalous dimension of the field itself defined in (1.10)

Relevant, Irrelevant and Marginal

Couplings in the neighbourhood of a fixed point flow as in (1.22) and are accordingly classified in the following way:

(i) If a coupling has $\Delta_i < d$ the flow diverges away from the fixed point into the IR (μ decreasing) and is known as a *relevant* coupling.
(ii) If $\Delta_i > d$ the coupling flows back into the fixed point and is known as an *irrelevant* coupling.
(iii) The case $\Delta_i = d$ is a marginal coupling for which one has to go to higher order to find out the behaviour. If, due to the higher order terms, a coupling diverges away/converges towards the fixed point it is *marginally relevant/irrelevant*. The final possibility is that the coupling does not run to all orders. In this case it is a *truly marginal* coupling and this implies that the original fixed point is actually part of a whole line of fixed points.

In a CFT the Green's functions are covariant under scale transformations and this provides non-trivial constraints. As an example, consider the 2-point Green's function $\Gamma^{(2)}(x) = \langle \phi(x)\phi(0) \rangle$. This satisfies the more general RG equation (1.12)

$$Z(\mu)^{-1} \Gamma^{(2)}(x; \mu, g_i(\mu)) = Z(\mu')^{-1} \Gamma^{(2)}(x; \mu', g_i(\mu')) . \tag{1.24}$$

At a fixed point $g_i(\mu) = g_i(\mu') = g_i^*$ and $Z(\mu) = (\mu'/\mu)^{2\gamma_\phi^*} Z(\mu')$, where $\gamma_\phi^* = \gamma_\phi(g_i^*)$. Then using dimensional analysis we must have

$$\Gamma^{(2)}(x; \mu, g_i^*) = \mu^{2d_\phi} \mathscr{G}(x\mu) , \tag{1.25}$$

where d_ϕ is the classical dimension of the field ϕ. Substituting into the RG equation (1.24) allows us to solve for the unknown function $\mathscr{G}$ (up to an overall multiplicative constant) yielding

$$\Gamma^{(2)}(x; \mu, g_i^*) = \frac{c}{\mu^{2\gamma_\phi^*} x^{2d_\phi + 2\gamma_\phi^*}} \propto \frac{1}{x^{2\Delta_\phi}} \tag{1.26}$$

where c is a constant. This displays the typical power-law behaviour characteristic of correlation functions in a CFT.

When we follow an RG flow backwards towards the UV (against the direction of the arrows in the figures below) all physical particle masses relative to cut off m_{phys}/μ decrease[5] so the trajectory of a theory with a mass gap must approach the critical surface. In typical cases, with or without a mass-gap, the trajectory diverges off to

---

[5] This is true in both versions of the RG equations (1.2), where m_{phys} is fixed and μ decreases, and (1.3), where μ is fixed and m_{phys} increases.

infinity for finite μ. However, with some fine tuning, the trajectory can approach a fixed point lying on the critical surface in the limit $\mu \to \infty$. Below we show the RG flows around a fixed point with 2 irrelevant directions and 1 relevant direction.

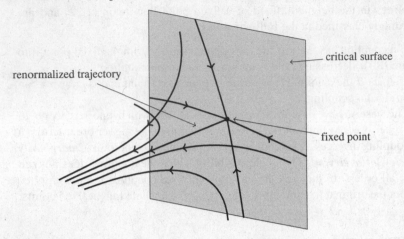

Notice that the flows lying off the critical surface naturally focus onto the "renormalized trajectory", which is defined as the flow that comes out of the fixed point. The focusing effect, along with the fact that there are only a small number of relevant directions, leads to the property of *Universality* which is a key feature of RG flows. Universality also arises for flows starting on the critical surface since in this case they flow into the fixed point.

Universality

CFTs only have a finite—and usually small—number of relevant couplings. This means that the domain-of-attraction, the number of irrelevant couplings, of a fixed point (the set of all points in theory-space that flow into a fixed point) is infinite dimensional. This also means that RG flows of a theory lying off the critical surface strongly focus onto finite dimensional subspaces parametrized by the relevant couplings of a fixed point, as in the figure above with one relevant coupling. The implication of this is that the behaviour of theories in the IR is determined by only a small number of relevant couplings and not by the infinite set of couplings $\{g_i\}$. This means that IR behaviour of a theory lie in a small set of "universality classes" which are associated to the domains of attraction of one, or possibly several, fixed points.

1.4 The Continuum Limit

The RG is directly relevant to the problem of taking a continuum limit of a QFT. This is the process of taking the cut off from its original finite value μ to infinity whilst keeping the physics at any momentum less than the original μ fixed. Whether such a limit exists is a highly non-trivial issue. Notice that taking a continuum limit involves the inverse RG flow, that is $g_i(\mu)$ with μ increasing.

The RG Equation (1.2) shows how this can be achieved. We can send $\mu \to \infty$ as long as the UV limit of $g_i(\mu)$ is suitably well-defined, which in practice means that $g_i(\infty)$ is a fixed-point of the RG. The resulting $g_i(\mu)$ which emanates from the fixed point is known as a "renormalized trajectory" and it provides a definition of the theory on all length scales. Searching for a renormalized trajectory would seem to involve searching for a needle in an infinite haystack since it requires fine-tuning an infinite set of couplings. Fortunately, however, universality comes to our rescue: we do not need to actually sit precisely on the renormalized trajectory in order to define a continuum theory. All that is required is a one-parameter set of theories defined with cut off μ' and with couplings $g_i = \tilde{g}_i(\mu')$ (which is generally *not* an RG flow since this would require sitting precisely on the renormalized trajectory) for which $\tilde{g}_i(\infty)$ lies on the critical surface in the domain of attraction of the fixed point associated to the CFT, as illustrated below.

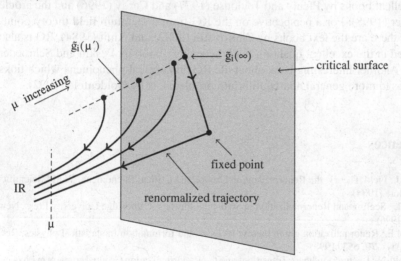

The limit $\mu' \to \infty$ is defined in such a way that the IR physics at the original cut-off scale μ is fixed. In particular, the number of parameters that must be specified in order to take a continuum limit, i.e. which fix the IR physics, equals the number of relevant couplings of the CFT. However, both the way that relevant couplings are fixed at the scale μ and the way that the values of the irrelevant couplings $\tilde{g}_i(\mu')$ behave as $\mu' \to \infty$ can be chosen in many different ways. Consequently there are many ways to take a continuum limit, or RG "schemes", which all lead to the same continuum theory. In particular, in particle physics we can exploit this freedom to take very simple forms for the action with a small number of terms (equal to the number of relevant coupling of the UV CFT). However, at the same time it allows us to describe the same QFT by using a lattice cut off with an action that seems very different from the continuum action.

Of course one could argue that there is no real reason why we should take a continuum limit in particle physics. As long as we keep the cut off just above the momentum scale of interest, then our effective theories are adequate. However, it is an interesting question to ask how far into the UV we can push a given effective theory before it becomes inconsistent because this gives us a window into more fundamental physics at shorter distance scales.

Bibliographical Notes

The version of the RG as pioneered by Wilson is described by him in the reviews (Wilson and Kogut 1974; Wilson 1983). The latter work, which is a copy of Wilson's Nobel Prize Lecture in 1982, has an excellent summary and list of references of the history of renormalization and the RG in both statistical physics and quantum field theory, as well as a beautiful and intuitive explanation of the RG. There are many good introductions to the RG from a more statistical mechanics perspective including the excellent books by Pfeuty and Toulouse (1977) and Cardy (1996) and the article by Fisher (1998). For a perspective on the RG from a quantum field theory point-of-view there are the text books by Zinn-Justin (2002) and Amit (1984). RG is also discussed in the excellent quantum field theory text-book by Peskin and Schroeder (1995). Another interesting book about the RG and critical phenomena which links RG ideas to more general non-equilibrium problems, is by Goldenfeld (1992).

References

Amit, D.J.: Field Theory, the Renormalization Group, and Critical Phenomena. World Scientific, Singapore (1984)

Cardy, J.L.: Scaling and Renormalization in Statistical Physics. Cambridge University Press, New York (1996)

Fisher, M.E.: Renormalization group theory: its basis and formulation in statistical physics. Rev. Mod. Phys. **70**, 653 (1998)

Goldenfeld, N.: Lectures on Phase Transitions and the Renormalization Group, Frontiers in Physics, vol. 85. Addison-Wesley, New York (1992)

Peskin, M.E., Schroeder, D.V.: An Introduction to Quantum Field Theory. Westview Press, New York (1995)

Pfeuty, P., Toulouse, G.: Introduction to the Renormalization Group and Critical Phenomena. Wiley, Chichester (1977)

Wilson, K.G., Kogut, J.B.: The renormalization group and the epsilon expansion. Phys. Rept. **12**, 75 (1974)

Wilson, K.G.: The renormalization group and critical phenomena. Rev. Mod. Phys. **55**, 583 (1983)

Zinn-Justin, J.: Quantum Field Theory and Critical Phenomena. Clarendon Press, Oxford (2002)

Chapter 2
Scalar Field Theories

In this chapter we put the RG concept to work on the simplest QFT based on a single real scalar field. This will illustrate how the conceptually simple idea of RG actually becomes challenging to implement in practice.

In particle physics we often write down simple actions like[1]

$$S[\phi] = \int d^d x \left(-\frac{1}{2} \partial_\mu \phi \partial^\mu \phi - \frac{1}{2} m^2 \phi^2 - \frac{1}{4!} \lambda \phi^4 \right); \tag{2.1}$$

however, in the spirit of RG we should—at least to start with—allow all possible terms consistent with space-time symmetries. In the case of a scalar field this involves all powers of the field and its derivatives contracted in a Lorentz invariant way. For simplicity we shall restrict to operators even under the symmetry $\phi \to -\phi$. This is an example of using a symmetry to restrict the possible couplings, the important point being that the symmetry is respected by RG flow. Simple scaling analysis shows that a composite operator $\mathcal{O}$ containing p derivatives and $2n$ powers of the field, schematically $\partial^p \phi^{2n}$, has classical mass dimension[2]

$$d_{\mathcal{O}} = n(d-2) + p. \tag{2.2}$$

Even at the classical level we see that the number of relevant/marginal couplings— those with $d_{\mathcal{O}} \leq d$—is small. The table below classifies some of the operators as relevant, marginal and irrelevant according to the dimension of space-time.

---

[1] In our notation, the scalar product in Minkowski space is $a_\mu b^\mu = \eta_{\mu\nu} a^\mu b^\nu = -a^0 b^0 + \mathbf{a} \cdot \mathbf{b}$.
[2] Note that the mass dimension of the field itself is fixed by the kinetic term to be $\frac{d-2}{2}$.

T. J. Hollowood, *Renormalization Group and Fixed Points*, SpringerBriefs in Physics, DOI: 10.1007/978-3-642-36312-2_2, © The Author(s) 2013

$\mathcal{O}$	$d > 4$	$d = 4$	$d = 3$	$d = 2$
ϕ^2	rel	rel	rel	rel
ϕ^4	irrel	marg	rel	rel
ϕ^6	irrel	irrel	marg	rel
ϕ^{2n}	irrel	irrel	irrel	rel
$(\partial_\mu \phi)^2$	marg	marg	marg	marg
$\phi^{2n}(\partial_\mu \phi)^2$	irrel	irrel	irrel	marg

The classical scaling suggests that, at least in dimensions $d > 2$, we only need to keep track of the kinetic term along with a completely general potential energy term, that is

$$\mathcal{L} = -\frac{1}{2}(\partial_\mu \phi)^2 - V(\phi), \tag{2.3}$$

where we take

$$V(\phi) = \sum_n \mu^{d-n(d-2)} \frac{g_{2n}}{(2n)!} \phi^{2n}. \tag{2.4}$$

In the above, we have used the powers of the cut off μ in order to have dimensionless couplings g_{2n}.

2.1 Finding the RG Flow

Now we come to crux of the problem, that of finding the RG flow. In order to do this we must apply the RG Eq. (1.4) to the Wilsonian Effective Action $S[Z(\mu)^{1/2}\varphi; \mu, g_i(\mu)]$, defined for the theory with cut off μ in such a way that the observables on momentum scales below the cut off are fixed as μ is varied.

Before we can describe how to relate the theories with different cut offs we must first settle on a particular cut-off procedure. The most basic and conceptually simple way to regularize a scalar field theory is to introduce a sharp momentum cut off on the Fourier modes after Wick rotation to Euclidean space. In Euclidean space the Lagrangian (2.3) has the form

$$\mathcal{L}_E = \frac{1}{2}(\partial_\mu \phi)^2 + V(\phi) \tag{2.5}$$

with $S_E = \int d^d x \, \mathcal{L}_E$ and the functional integral becomes $\int [d\phi] \exp(-S_E)$.[3] The momentum cut off involves Fourier transforming the field

---

[3] We take it as established fact that one can transform between the Minkowski and Euclidean versions of the theory without difficulty. In our conventions, the Wick rotation involves $\eta_{\mu\nu}a^\mu b^\nu = -a_0 b_0 + \mathbf{a} \cdot \mathbf{b} \to a_\mu b_\mu = a_0 b_0 + \mathbf{a} \cdot \mathbf{b}$. In Euclidean space the functional integral $\int [d\phi] e^{-S_E[\phi]}$ can be interpreted as a probability measure (when properly normalized) on the field configuration space. This is why Euclidean QFT is intimately related to systems in statistical physics. In the following

$$\phi(x) = \int \frac{d^d p}{(2\pi)^d} \tilde{\phi}(p) e^{ip \cdot x} \tag{2.6}$$

and then limiting the momentum vector by means of a sharp cut off $|p| \leq \mu$. The resulting theory is now manifestly UV finite since the momenta in loops are never taken all the way to infinity. In addition, we have a very concrete way of performing the RG transformation. Namely, we split the field ϕ defined with cut off μ' into two pieces

$$\phi = \varphi + \hat{\phi}, \tag{2.7}$$

where φ has the modes with $|p| \leq \mu$, while $\hat{\phi}$ has the modes in the interval $\mu \leq |p| \leq \mu'$:

$$\phi(x) = \int_{|p| \leq \mu'} \frac{d^d p}{(2\pi)^d} \tilde{\phi}(p) e^{ip \cdot x}$$

$$= \underbrace{\int_{|p| \leq \mu} \frac{d^d p}{(2\pi)^d} \tilde{\phi}(p) e^{ip \cdot x}}_{\varphi(x)} + \underbrace{\int_{\mu \leq |p| \leq \mu'} \frac{d^d p}{(2\pi)^d} \tilde{\phi}(p) e^{ip \cdot x}}_{\hat{\phi}(x)} . \tag{2.8}$$

In order to extract the beta functions it is sufficient to consider the infinitesimal transformation with $\mu' = \mu + \delta\mu$. We can then obtain the RG flow by considering how the action changes after we integrate out the Fourier modes $\hat{\phi}$; so concretely,

$$\exp\left(-S[Z(\mu)^{1/2}\varphi; \mu, g_{2n}(\mu)]\right)$$

$$= \int [d\hat{\phi}] \exp\left(-S[\varphi + \hat{\phi}; \mu + \delta\mu, g_{2n}(\mu + \delta\mu)]\right) . \tag{2.9}$$

On both sides, we have the Wilsonian Effective Action defined at the scales μ and $\mu + \delta\mu$, respectively. Notice that without-loss-of-generality we have scaled the field on the right-hand side so that $Z(\mu + \delta\mu) = 1$.

Expanding the action on the right-hand side in powers of $\hat{\phi}$[4]:

$$S[\varphi + \hat{\phi}] = S[\varphi] + \int d^d x \left(\frac{1}{2}(\partial_\mu \hat{\phi})^2 + \frac{1}{2} V''(\varphi)\hat{\phi}^2 + \frac{1}{6} V'''(\varphi)\hat{\phi}^3 + \cdots\right). \tag{2.10}$$

Note that the cross term $\int d^d x\, \partial_\mu \varphi \partial_\mu \hat{\phi}$ vanishes, a fact that becomes obvious when written in terms of momentum space modes. The non-trivial part of the problem is

---

we shall not show the subscript E for "Euclidean" since the context will dictate whether we are working in Minkowski or Euclidean space.

[4] In writing the action we have ignored the term $\hat{\phi} V'(\varphi)$ since its presence does not affect the effective potential that we calculate below. To appreciate this note that we can effectively take φ to be constant and then shift it to lie at the minimum of $V(\varphi)$.

to integrate over the modes $\hat{\phi}$. One way to do this is to work in terms of Feynman diagrams:

Feynman Diagram Interpretation

The terms that one gets in the effective action, after integrating out $\hat{\phi}$, can be interpreted in terms of Feynman diagrams with only $\hat{\phi}$ on internal lines contributing a propagator

$$\frac{1}{p^2 + g_2\mu^2} \quad \sim \quad \hat{\phi} \;\text{-------}\; \hat{\phi}$$

and with only φ on external lines (but with no propagators). The vertices are provided by expanding the potential $V(\varphi + \hat{\phi})$ in powers of φ and $\hat{\phi}$. As an example, there is a vertex of the form

$$\frac{g_6}{4!2!}\varphi^4\hat{\phi}^2 \quad \sim \quad$$

Each loop involves an integral over the momentum of $\hat{\phi}$ which lies in a shell between radii μ and μ' in momentum space:

$$\int_{\mu \leq |p| \leq \mu'} \frac{d^d p}{(2\pi)^d}\, f(p). \tag{2.11}$$

If we are only interested in an infinitesimal RG transformation $\mu' = \mu + \delta\mu$ then the integrals over loop momenta (2.11) become much simpler:

$$\int_{\mu \leq |p| \leq \mu+\delta\mu} \frac{d^d p}{(2\pi)^d} f(p) = \frac{\mu^{d-1}\delta\mu}{(2\pi)^d} \int d^{d-1}\hat{\Omega}\, f(\mu\hat{\Omega}), \tag{2.12}$$

where $\hat{\Omega}$ is a unit d vector to be integrated over a unit $d-1$-dimensional sphere S^{d-1}. Now comes the key part of the calculation: since each loop integral brings along a factor of $\delta\mu$, to linear order in $\delta\mu$, only one loop diagrams are needed. Since the dependence on the momentum in the loop can only be via the invariant $p^2 = \mu^2$, the integral over the solid angle $\hat{\Omega}$ just yields a constant $\text{Vol}(S^{d-1})$, the volume of a $d-1$-dimensional sphere. Even with this huge simplification, we still have the non-trivial combinatorial problem of summing over an infinite set of one-loop diagrams one of which we show as an example below:

$$\sim \quad \frac{g_6 \cdot g_4 \cdot g_6 \cdot g_8}{2^5 (4!)^2 6!} \frac{\mu^{d-1} \delta\mu \, \mathrm{Vol}(S^{d-1})}{(2\pi)^d} \frac{1}{(\mu^2 + g_2 \mu^2)^4}$$

Fortunately, there is a simple way to sum all such one-loop diagrams in one go. The trick is to keep only the terms quadratic in $\hat\phi$ in (2.10),

$$S_2[\hat\phi] = \int d^d x \left(\frac{1}{2}(\partial_\mu \hat\phi)^2 + \frac{1}{2} V''(\varphi)\hat\phi^2 \right) \tag{2.13}$$

and then perform the resulting Gaussian integral over $\hat\phi$

$$e^{-\delta S} = \int [d\hat\phi] e^{-S_2[\hat\phi]}. \tag{2.14}$$

In order to extract the RG transformation, we identify the change in the Wilsonian effective action as

$$S[\varphi; \mu, g_{2n}(\mu)] - S[\varphi; \mu + \delta\mu, g_{2n}(\mu + \delta\mu)] = \delta S. \tag{2.15}$$

In the present case there is no wave-function renormalization.[5] Since we are ultimately interested in the effective potential only, we can temporarily assume that φ is a constant. In the that case, denoting the Fourier transform of $\hat\phi(x)$ as $\tilde\phi(p)$, we have

$$S_2 = (\mu^2 + V''(\varphi)) \cdot \frac{\mu^{d-1}\delta\mu}{(2\pi)^d} \int d^{d-1}\hat\Omega \, \tilde\phi(\mu\hat\Omega)\tilde\phi(\mu\hat\Omega), \tag{2.16}$$

using the fact that $p_\nu p_\nu = \mu^2$ for the modes $\tilde\phi$. Hence, performing the Gaussian integral over the modes $\tilde\phi$ yields the result

$$e^{-\delta S} = C \left(\frac{\pi}{\mu^2 + V''(\varphi)} \right)^{\mathcal{N}/2}, \tag{2.17}$$

where $\mathcal{N}$ is the number of modes in the momentum shell. One way to count the modes is to introduce an infra-red cut off by defining the theory in a very large box of size L and imposing periodic boundary conditions on the field. This has the effect of quantizing the components of momenta as $p_\mu = 2\pi n_\mu / L$, for integers n_μ, in units $\hbar = 1$. We see that, effectively, there is one mode per volume $(2\pi)^d$ in Euclidean phase space. Therefore, assuming that L is sufficiently large, and denoting the volume of space-time as $\mathcal{V} = L^d$, the number of modes in the shell in momentum space is

---

[5] This is because in this theory at one-loop order the only Feynman diagram with two external legs has no external momentum running in the loop, and so cannot give rise to a term proportional to $(\partial_\mu \varphi)^2$.

$$\mathcal{N} = \frac{\text{Vol}(S^{d-1})}{(2\pi)^d} \, \mu^{d-1} \, \delta\mu \, \mathcal{V}. \tag{2.18}$$

In this case, we can write the contribution in (2.17), up to an unimportant overall constant, as

$$\exp\left(-\delta S\right) = \exp\left(-a\mu^{d-1}\delta\mu \int d^d x \, \log(\mu^2 + V''(\varphi))\right), \tag{2.19}$$

where we have defined $a = \text{Vol}(S^{d-1})/(2(2\pi)^d) = 2^{-d}\pi^{-d/2}/\Gamma(\frac{d}{2})$. Note that we have replaced the volume factor $\mathcal{V}$ by the integral over space-time which allows us to remove the temporary assumption that φ is constant. Hence, we can identify the change in the Wilsonian effective action as

$$S[\varphi; \mu, g_{2n}(\mu)] - S[\varphi; \mu + \delta\mu, g_{2n}(\mu + \delta\mu)]$$
$$= a\mu^{d-1} \int d^d x \, \log(\mu^2 + V''(\varphi)) \, \delta\mu. \tag{2.20}$$

Expanding the right-hand side in powers of φ, allows us to extract the beta functions of the couplings directly:

$$\mu\frac{dg_{2n}}{d\mu} = (n(d-2) - d) \, g_{2n} - a\mu^{n(d-2)} \frac{d^{2n}}{d\varphi^{2n}} \log(\mu^2 + V''(\varphi))\Big|_{\varphi=0}. \tag{2.21}$$

If we expand in powers of the coupling constants, the contributions on the right-hand side can be identified with individual one-loop diagrams like the one shown previously.

From (2.21), the first few beta functions in the hierarchy are

$$\mu\frac{dg_2}{d\mu} = -2g_2 - \frac{ag_4}{1 + g_2},$$

$$\mu\frac{dg_4}{d\mu} = (d-4)g_4 + \frac{3ag_4^2}{(1 + g_2)^2} - \frac{ag_6}{1 + g_2},$$

$$\mu\frac{dg_6}{d\mu} = (2d-6)g_6 - \frac{30ag_4^3}{(1 + g_2)^3} + \frac{15ag_4g_6}{(1 + g_2)^2} - \frac{ag_8}{1 + g_2}. \tag{2.22}$$

Notice that the quantum contributions involve inverse powers of the factor $1 + g_2$ which physically is $m^2/\mu^2 + 1$, where m is the mass of the field. These factors arise from the propagators of the modes in the loop. So when $m \gg \mu$, the quantum terms are suppressed as one would expect on the basis of decoupling.[6]

---

[6] Decoupling expresses the intuition that a particle of mass m cannot directly affect the physics on distance scales $\gg m^{-1}$. For instance, the potential due to the exchange of massive particle in four dimensions is $\sim e^{-mr}/r$. This is exponentially suppressed on distances scales $\gg m^{-1}$. In Chap. 4 we will make the notion of decoupling more precise.

The part of the Wilson effective action that depends just on the field and not its derivatives in the IR limit as $\mu \to 0$ is known as the *effective potential* $V_{\text{eff}}(\varphi)$. It plays an important rôle since its minima determine the possible vacuum states of the theory. In the present approximation

$$V_{\text{eff}}(\varphi) = \lim_{\mu \to 0} \sum_n \mu^{d-n(d-2)} \frac{g_{2n}(\mu)}{(2n)!} \varphi^{2n}. \tag{2.23}$$

Vacuum Expectation Values

In a scalar QFT, the field can develop a non-trivial Vacuum Expectation Value (VEV) $\langle \phi \rangle \neq 0$. This possibility is determined by finding the global minima of the effective potential $\langle \phi \rangle = \varphi$. It can be shown that the effective potential defined in terms of the Wilsonian effective action is equal to the effective potential extracted from the more familiar 1-Particle Irreducible (1-PI) effective action defined in perturbation theory—at least for a QFT with a mass gap. A VEV develops when the effective potential develops minima away from the origin as in the example

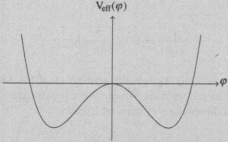

Since we started with a theory symmetric under $\varphi \to -\varphi$, there will necessarily be two possible vacuum states with opposite values of the VEV $\langle \phi \rangle$. The theory must choose one or the other and so we say that the symmetry $\varphi \to -\varphi$ is *spontaneously broken*. The reason why spontaneous symmetry breaking can occur is that the constant mode of a scalar field is not part of the variables that appear in the measure of the functional integral $\int [d\phi]$. The rôle of the constant mode is to act as a boundary condition on the scalar field at spatial infinity. However, this is only true in space-time dimensions $d > 2$: in $d = 2$ small fluctuations can change the field at spatial infinity and so for consistency one must also integrate over the zero mode as well and as a consequence, spontaneous symmetry breaking of a *continuous* symmetry cannot occur. This is the statement of the Mermin-Wagner Theorem. In our case $\varphi \to -\varphi$ is a discrete symmetry and so escapes the implications of the theorem and can still be spontaneously broken even in $d = 2$.

2.2 Mapping the Space of Flows

The beta functions allow us to map-out the RG flow on theory space. The first thing to do is to find the RG fixed points corresponding to the CFTs. The "Gaussian fixed point" is the trivial fixed point where all the couplings vanish $g_{2n} = 0$. Linearizing around this point, the beta-functions are

$$\mu \frac{dg_{2n}}{d\mu} = (n(d-2) - d)\, g_{2n} - a g_{2n+2}. \tag{2.24}$$

So the scaling dimensions at the Gaussian fixed point are simply the classical dimensions $\Delta_{2n} = d_{2n} = n(d-2)$, i.e. the anomalous dimensions vanish, although the couplings that diagonalize the matrix of scaling dimensions σ_{2n} are not precisely equal to g_{2n} due to the second term in (2.24). In particular, $\sigma_2 = g_2$ is always relevant, $\sigma_4 = g_4 + a g_2/(2-d)$ is relevant for $d < 4$, irrelevant for $d > 4$ and marginally irrelevant for $d = 4$. In this latter case we need to go beyond the linear approximation. Since g_6 is irrelevant in $d = 4$, we shall ignore it, and using $a = 1/16\pi^2$ we have

$$\mu \frac{dg_4}{d\mu} = \frac{3}{16\pi^2} g_4^2, \tag{2.25}$$

whose solution is

$$\frac{1}{g_4(\mu)} = C - \frac{3}{16\pi^2} \log \mu. \tag{2.26}$$

This shows that g_4 is actually *marginally irrelevant* at the Gaussian fixed point because it gets smaller as μ decreases. We can write the integration constant in terms of a parameter Λ with dimensions of mass as follows:

$$g_4(\mu) = \frac{16\pi^2}{3\log(\Lambda/\mu)}, \tag{2.27}$$

with $\mu < \Lambda$. This is an example of *dimensional transmutation*, where the freedom to specify a dimensionless coupling g_4 in the action turns into a quantity Λ with unit mass dimension in the quantum theory. Notice that Λ is the momentum scale at which the running coupling $g_4(\mu)$ diverges. Clearly, perturbation theory will break down as this scale is approached as we discuss in Chap. 3.

To find other non-trivial fixed points is difficult, but one way we can make progress is to work perturbatively in the couplings. In order to do this we have to do something that looks counter-intuitive and consider the RG flow equations in arbitrary non-integer dimension d regarding $\varepsilon = 4 - d$ as a small parameter. Hopefully, what is established for small ε will be qualitatively true for in the physical cases where $\varepsilon = 1, 2$, etc. Proceeding in this way, we find a new non-trivial fixed point known as the Wilson-Fisher fixed point at

$$g_2^* = -\frac{1}{6}\varepsilon + \cdots, \quad g_4^* = \frac{1}{3a}\varepsilon + \cdots, \quad g_{2n>4}^* \sim \varepsilon^n + \cdots \tag{2.28}$$

In particular, the Wilson-Fisher fixed point is only physically acceptable if $\varepsilon > 0$, or $d < 4$, since otherwise the couplings g_{2n}^* are all negative and the potential of the theory would not be bounded from below leading to an instability. In the neighbourhood of the fixed point in the (g_2, g_4) subspace, to linear order in ε, we have

$$\mu \frac{d}{d\mu} \begin{pmatrix} \delta g_2 \\ \delta g_4 \end{pmatrix} = \begin{pmatrix} \varepsilon/3 - 2 & -b(1 + \varepsilon/6) \\ 0 & \varepsilon \end{pmatrix} \begin{pmatrix} \delta g_2 \\ \delta g_4 \end{pmatrix} \tag{2.29}$$

with

$$b = \frac{1}{16\pi^2} + \frac{\varepsilon}{32\pi^2}(1 - \gamma_E + \log 4\pi) + \mathcal{O}(\varepsilon^2). \tag{2.30}$$

So the scaling dimensions of the associated operators and the associated couplings at the Wilson-Fisher fixed point are

$$\Delta_2 = 2 - \frac{2\varepsilon}{3}, \quad \sigma_2 = \delta g_2,$$
$$\Delta_4 = 4, \quad \sigma_4 = 2(3 + \varepsilon)\delta g_4 - b(3 + \varepsilon/2)\delta g_2. \tag{2.31}$$

Therefore at this fixed point only the mass coupling σ_2 is relevant.

The flows in the (g_2, g_4) subspace of scalar QFT for small $\varepsilon > 0$ are shown below:

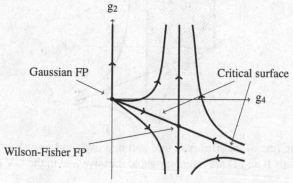

The Gaussian and Wilson-Fisher fixed points are shown and since all the other couplings are irrelevant we only show the flows in the (g_2, g_4) subspace. It is important to realize, as is clear from (2.28), that the irrelevant couplings do not vanish on this subspace. Notice that the critical surface intersects this subspace in the line that joins the two fixed points as shown.

Although we have only proved the existence of the Wilson-Fisher fixed point for small ε, it is known to exist in both $d = 3$ and $d = 2$. In the language of statistical physics it lies in the universality class of the Ising Model.[7] What our simple analysis

---

[7] The Ising Model is a statistical model defined on a square lattice with spins $\sigma_i \in \{+1, -1\}$ at each site and with an energy (which we identify with the Euclidean action)

fails to show is that in $d = 2$ there are actually an infinite sequence of additional fixed points.[8]

Now that we have a qualitative picture of the RG flows, it is possible to describe the possible continuum limits of scalar field theories:

$d = 4$: In this case, only the Gaussian fixed point exists and this fixed point only has one relevant direction; namely, the mass coupling g_2. Hence there is a single renormalized trajectory on which $g_2(\mu) = (\mu'/\mu)^2 g_2(\mu')$ while all the other couplings vanish. This renormalized trajectory describes the free massive scalar field. If we sit precisely at the fixed point we have a free massless scalar field. In particular, according to our crude analysis there is no interacting continuum theory in $d = 4$. In order to have an interaction the cut off must be kept finite. It turns out that this conclusion is backed up in more sophisticated analyses. This important fact, known as "triviality", has important implications for the Higgs sector of the standard model as we describe in Chap. 3.

$d = 3$: Assuming the existence of the Wilson-Fisher fixed point there are two fixed points and a two-dimensional space of renormalized trajectories parametrized by the couplings g_2 and g_4 on which g_{2n}, $n > 2$ have some values fixed by g_2 and g_4. In particular, if we parametrize our continuum theories by the values of g_2 and g_4 then they are limited to the regions shown below:

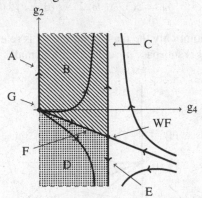

In particular, the line of theories **A** is free and massive (and must have $g_2 > 0$); the theories in regions **B** and **D** are interacting and massive and in the UV becomes non-

---

(Footnote 7 continued)

$$\mathscr{E} = -\frac{1}{T} \sum_{(i,j)} \sigma_i \sigma_j. \tag{2.32}$$

The sum is over all nearest-neighbour pairs (i, j) and T is the temperature. Notice that at low temperatures, the action/energy favours alignment of all the spins, while at high temperatures thermal fluctuations are large and the long-range order is destroyed. This can be viewed as a competition between energy and entropy. There is a 2nd order phase transition at a critical temperature $T = T_c$ at which there are long-distance power-law correlations. This critical point is in the same universality class as the Wilson-Fisher fixed point.

[8] In $d = 2$ there are powerful exact methods for analyzing CFTs because in $d = 2$ the conformal group is infinite dimensional as it consists of any holomorphic transformation $t \pm x \rightarrow f_\pm(t \pm x)$.

interacting, or free, since the trajectories originate from the Gaussian fixed point. The continuum theories are parametrized by two couplings, a mass scale and an interaction strength. In case **D**, $g_2 < 0$ and the field has a VEV; the line of theories **C** and **E** describe massive interacting theories that become the Ising model CFT in the UV. Furthermore, case **E** has $g_2 < 0$ and a VEV; the line of theories **F** describe a massless interacting theory that interpolates between a free theory in the UV and the Ising Model CFT in the IR; the point **G** is a free massless theory (the Gaussian fixed point); and the point **WF** is the Ising model CFT (the Wilson-Fisher fixed point). Any points to the right of the line **C–E** do not have continuum limits.

RG Crossover

The RG flows in $d = 3$ illustrate the notion of an RG *crossover*. Consider the two theories associated to the RG trajectories that we denoted as **B** and **D** shown again below:

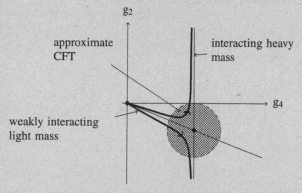

These trajectories issue from the Gaussian fixed point and in the far UV the theories become free, a property more common in gauge theories known as *asymptotic freedom*, studied in Chap. 4. At intermediate energy scales the trajectories pass close to the Wilson-Fisher fixed point (the dotted region) and the spectrum consists of a light interacting scalar m with $m \ll \mu$. The theory here is approximately conformally invariant with the physics determined approximately by the Wilson-Fisher CFT. However, as μ runs down to and beyond m, the RG trajectory veers away the Wilson-Fisher fixed-point and the theory becomes interacting and massive.

Bibliographical Notes

The implementation of the Wilsonian RG in scalar field theories with a sharp momentum cut off is a well-studied problem. The approximation of the full effective action by the kinetic term and effective potential is known as the *local potential approximation* and was studied originally by Wegner and Houghton (1973). The problem is studied in Chap. 12 of Peskin and Schroeder (1995). Approaches that go beyond this are known as the "exact", "functional" or "continuous" RG and have been developed by many authors including, for example, Polchinski (1984), Morris (1996) and Polonyi (2003).

References

Morris, T.R.: Nucl. Phys. B **458**, 477 (1996). [hep-th/9508017]
Peskin, M.E., Schroeder, D.V.: An Introduction to Quantum Field Theory. Addison-Wesley, Reading (1995)
Polchinski, J.: Nucl. Phys. B **231**, 269 (1984)
Polonyi, J.: Central Eur. J. Phys. **1**, 1 (2003). [hep-th/0110026]
Wegner, F.J., Houghton, A.: Phys. Rev. A **8**, 401 (1973)

Chapter 3
RG and Perturbation Theory

In this chapter we consider how to implement RG ideas in the context of perturbation theory, and in a way which will be easier to generalize to other theories including gauge theories which we treat in the following chapter.

We have seen that it is the remarkable focussing properties of the RG flows and the consequent universality that allows one to formulate theories in terms of simple actions. The action only needs to include a kinetic term and the relevant interactions; in particular, all the irrelevant couplings can be taken to vanish. The argument runs as follows: in order to take a continuum limit, we need to let the "bare couplings" $g_i = \tilde{g}_i(\mu')$ depend on the cut off μ' in such a way that the phenomena at a physically relevant scale $\mu < \mu'$ remains fixed. This is guaranteed if we use the RG equation (1.2) and take $\mu' \to \infty$ with the couplings $g_i = g_i(\mu')$ following the RG flow into the UV along a renormalized trajectory. However, this would mean fine tuning all the irrelevant couplings. As we described in Chap. 1, there is no need to do this. It is sufficient to keep only the relevant couplings $g_2(\mu')$ and $g_4(\mu')$ in addition to the kinetic term and choose all the irrelevant couplings to vanish. The resulting behaviour of all the couplings, that is

$$
\begin{aligned}
\tilde{g}_2(\mu') &= g_2(\mu'), \\
\tilde{g}_4(\mu') &= g_4(\mu'), \\
\tilde{g}_{2n}(\mu') &= 0, \quad n > 2,
\end{aligned}
\tag{3.1}
$$

is not an RG flow but rather is a projection of the RG flow onto the finite-dimensional subspace (g_2, g_4) of the relevant couplings.

At low energies the flows from the subspace (3.1) are drawn to the renormalized trajectory and the true continuum theory is recovered. In particular, irrelevant couplings are generated but they are not free parameters in the low energy theory, on the contrary, they are fixed by the relevant couplings. The RG flow is illustrated below for the scalar field theory in dimension $d = 3$ with the magnitude of the irrelevant couplings exaggerated.

T. J. Hollowood, *Renormalization Group and Fixed Points*, SpringerBriefs in Physics, DOI: 10.1007/978-3-642-36312-2_3, © The Author(s) 2013

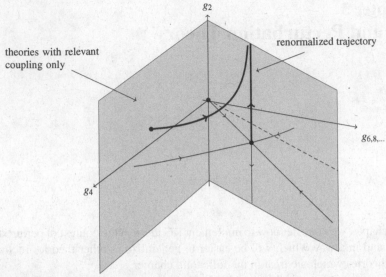

In many situations in QFT, perturbation theory is a powerful tool for uncovering the RG flows of the relevant couplings. In particular, even the one-loop approximation yields a wealth of insight. However, in the context of the RG, the use of perturbation poses an important question.

Perturbation Theory in Which Coupling?

According to the RG, the couplings flow with the scale of interest μ and this poses the question as to which coupling should be used as the perturbative parameter? For example, in $d = 4$, g_4 runs according to (2.26). So the coupling at the scale μ' is an infinite perturbative series of the coupling at μ:

$$g_4(\mu') = \frac{g_4(\mu)}{1 + \frac{3}{16\pi^2} g_4(\mu) \log \frac{\mu}{\mu'}} = g_4(\mu) - \frac{3}{16\pi^2} g_4(\mu)^2 \log \frac{\mu}{\mu'} + \cdots . \quad (3.2)$$

If we could completely sum all of the perturbative expansion in $g_4(\mu)$, then the resulting physical observables would, by the very definition of the RG, be independent of μ. However, since in real life it is inevitable that perturbation theory can only be performed to a given low order, it is clear that choosing $g_4(\mu)$ as the perturbative parameter will be different from choosing $g_4(\mu')$. The question is whether we really do have a choice for μ and can we try to be clever by choosing μ so that $g_4(\mu)$ is small and perturbation theory is more reliable? The answer is that as usual the appropriate choice for μ is dictated by the characteristic physical energy scale involved and we just have to hope that at this scale $g_4(\mu)$ is small. If we try to do perturbation at a non-physical scale, then any improvement we might make in having a smaller $g_4(\mu)$

is compensated by larger coefficients multiplying the powers of the coupling. We will see an example of how the physical scale dictates the choice of μ in a procedure known as *renormalization group improvement*.

The message is that we should perform the perturbation expansion in the coupling at the physically relevant energy scale μ, i.e. for the "renormalized coupling" $\lambda \equiv \lambda(\mu)$ (we choose to use λ and m rather than g_4 and g_2 in the following description) rather than the "bare coupling" $\lambda_b \equiv \lambda(\mu')$. To this end, we split up the bare Lagrangian into the "renormalized" part and the "counter-term" part:

$$\mathscr{L} = -\frac{1}{2}(\partial_\mu \phi_b)^2 - \frac{1}{2}m_b^2 \phi_b^2 - \frac{1}{4!}\lambda_b \phi_b^4 = \mathscr{L}_r + \mathscr{L}_{ct} \qquad (3.3)$$

where $\mathscr{L}_r$ has the same form as the bare Lagrangian, but with bare quantities replaced by renormalized ones (which we denote by ϕ, m and λ):

$$\mathscr{L}_r = -\frac{1}{2}(\partial_\mu \phi)^2 - \frac{1}{2}m^2 \phi^2 - \frac{1}{4!}\lambda \phi^4 \qquad (3.4)$$

and $\mathscr{L}_{ct}$, the counter-term Lagrangian, is given by

$$\mathscr{L}_{ct} = -\frac{1}{2}\delta Z(\partial_\mu \phi)^2 - \frac{1}{2}\delta m^2 \phi^2 - \frac{1}{4!}\delta \lambda \phi^4, \qquad (3.5)$$

such that the bare quantities are

$$m_b^2 = Z^{-1}(m^2 + \delta m^2), \quad \lambda_b = Z^{-2}(\lambda + \delta \lambda), \quad \phi_b = Z^{1/2}\phi, \qquad (3.6)$$

where $Z = 1 + \delta Z$. It is important to note that the counter-terms are *not* infinitesimals.

Perturbation theory can be performed in the renormalized coupling λ and the counter-terms found order-by-order in λ without prior knowledge of the RG flow. In this point-of-view, the counter-terms are chosen to cancel the divergences that occur in the limit $\mu' \to \infty$.[1] For instance, in $d < 4$, the only Feynman diagrams which are superficially divergent have 2 external legs:

---

[1] There is considerable freedom in choosing the way that the divergent terms are cancelled. These choices should be thought of as part of the regularization *scheme*. Different schemes correspond to different, but ultimately equivalent, parameterizations of the physical quantities in terms of the couplings in the action. This illustrates that we should not naïvely think of the g_i as physical quantities in themselves, rather they parametrize the physical quantities.

While in $d = 4$ all diagrams with 2 and 4 external lines are superficially divergent.

Superficial Degree of Divergence

The question of whether a given Feynman diagram is UV divergent can be partially addressed by calculating the *superficial degree of divergence* $\mathscr{D}$. This is the power of the overall momentum dependence of the diagram: each propagator contributes $\sim p^{-2}$ and each loop integral $\sim p^d$. It follows for the ϕ^4 theory that, for a diagram with L loops and E external lines,

$$\mathscr{D} = (d - 4)L + 4 - E. \tag{3.7}$$

Notice that when $\mathscr{D} \geq 0$ the diagram is divergent, however, the converse doesn't imply convergence since a sub-diagram may be divergent. When a theory only has superficial divergences in a finite number of diagrams it is called *super-renormalizable*.

While we could continue with the sharp momentum cut off, it is time to acknowledge its frailties. It has been a useful device to introduce the concept of RG flow, however, when we start to investigate gauge theories we discover that it is not obvious how to make the naïve sharp momentum cut off consistent with gauge invariance. The most common way to regularize a QFT in perturbation theory is motivated by formulating the theory in a space-time of arbitrary dimension d. In this *dimensional regularization* scheme the idea is to calculate perturbative contributions with $d = 4 - \varepsilon$ arbitrary and not necessarily integer. In fact after performing the loop integrals, the contributions will be analytic functions of ε. The divergences in the physical dimension, whether it be $d = 2, 3, 4$, show up as poles as $\varepsilon \to 2, 1, 0$. For example, in four dimensions we make the replacement

$$\int \frac{d^4 p}{(2\pi)^4} \longrightarrow \mu^{4-d} \int \frac{d^d p}{(2\pi)^d}, \tag{3.8}$$

where μ is a parameter, with unit mass dimension introduced to make sure that the momentum integrals have the correct mass dimension. We will see that in a subtraction scheme known as *minimal subtraction* it plays a rôle analogous to the Wilsonian cut-off scale μ, the scale at which the renormalized coupling are defined.

As an example consider the quadratically divergent one-loop diagram in $d = 4$:

which involves the loop integral

$$
\lambda \mu^{4-d} \int \frac{d^d p}{(2\pi)^d} \cdot \frac{1}{p^2 + m^2} = \lambda \mu^{4-d} \frac{\text{Vol}(S^{d-1})}{(2\pi)^d} \int_0^\infty \frac{p^{d-1} dp}{p^2 + m^2}
$$
$$
= \lambda 2^{-d} \pi^{-d/2} \mu^{4-d} m^{d-2} \Gamma(1 - \tfrac{d}{2}). \tag{3.9}
$$

In the limit $\varepsilon = 4 - d \to 0$ this equals

$$
-\frac{\lambda m^2}{8\pi^2 \varepsilon} + \frac{\lambda m^2}{16\pi^2} \Big[-1 + \gamma_E - \log \frac{4\pi \mu^2}{m^2} \Big] + \mathcal{O}(\varepsilon). \tag{3.10}
$$

The divergences can be removed by a mass counter-term of the form

$$
\delta m^2 = \frac{\lambda m^2}{16\pi^2 \varepsilon} + \frac{\lambda m^2}{32\pi^2} (-\gamma_E + \log 4\pi). \tag{3.11}
$$

Subtracting the divergence effectively cuts off the high momentum modes. As we have previously stated there is considerable freedom in removing the divergences. The choice (3.11) is known as *modified minimal subtraction*, denoted $\overline{\text{MS}}$. Here "minimal" refers to the singular part in (3.11), and "modified" refers to the second and third terms which are optional extras but part of the convention chosen in particle physics.

The other divergent diagram at one-loop is

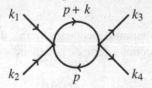

where $k = k_1 + k_2 = k_3 + k_4$. In this case the loop integral is logarithmically divergent in $d = 4$:

$$
\lambda^2 \mu^{4-d} \int \frac{d^d p}{(2\pi)^d} \cdot \frac{1}{p^2 + m^2} \cdot \frac{1}{(p + k)^2 + m^2}, \tag{3.12}
$$

where $k = k_1 + k_2$. Expanding in terms of the external momentum k, it is only the term of order k^0 which is divergent since every power of k effectively gives one less power of p for large p. The divergent piece is then

$$\lambda^2 \mu^{4-d} \int \frac{d^d p}{(2\pi)^d} \cdot \frac{1}{(p^2 + m^2)^2} = \lambda^2 \mu^{4-d} \frac{\text{Vol}(S^{d-1})}{(2\pi)^d} \int_0^\infty \frac{p^{d-1} dp}{(p^2 + m^2)^2}$$
$$= \lambda^2 2^{-d} \pi^{-d/2} \mu^{4-d} m^{d-4} \Gamma(2 - \tfrac{d}{2})$$
$$= \frac{\lambda^2}{8\pi^2 \varepsilon} + \frac{\lambda^2}{16\pi^2}\left[-\gamma_E + \log \frac{4\pi \mu^2}{m^2} \right] + \mathcal{O}(\varepsilon).$$
(3.13)

In minimal subtraction the divergence can be cancelled by the counter-term

$$\delta\lambda = \frac{3\lambda^2}{16\pi^2 \varepsilon} + \frac{3\lambda^2}{32\pi^2}(-\gamma_E + \log 4\pi). \tag{3.14}$$

Momentum Subtraction Scheme

The $\overline{\text{MS}}$ subtraction scheme is not the only way to remove the divergences in dimensional regularization. A more physically motivated scheme involves fixing the physical mass and physical coupling defined in terms of the propagator and 4-point Green's function at some characteristic space-like momentum scale $\tilde{\mu}$:

$$\Gamma^{(2)}(p_1, p_2) = \frac{1}{p_1^2 + m^2}\delta^{(4)}(p_1 + p_2),$$
$$\Gamma^{(4)}(p_1, p_2, p_3, p_4) = \lambda \delta^{(4)}(p_1 + p_2 + p_3 + p_4), \tag{3.15}$$

when $p_i^2 = \tilde{\mu}^2$.

In this kind of momentum-dependent scheme, the quantity $\tilde{\mu}$ plays the rôle of the RG scale. The counter-terms δg_i depend explicitly on $\tilde{\mu}$ and the beta functions are equal to

$$\tilde{\mu}\frac{d\delta g_i}{d\tilde{\mu}}. \tag{3.16}$$

We will see in Chap. 4, that the beta functions in this scheme are very different from the ones in the $\overline{\text{MS}}$ scheme, illustrating the potential dangers of identifying a running coupling directly with a physical coupling.

3.1 The Background Field Method

As mentioned above, the renormalized couplings will depend on the dimensional regularization mass scale μ in a way that is analogous to μ of the sharp momentum cut-off scheme. In fact, we can recover the spirit of Wilson's RG by using what is known as the *background field method*. The idea is to expand the field ϕ in the renormalized action in terms of a slowly varying background field φ and a more rapidly fluctuating part $\hat{\phi}$:

$$\phi = \varphi + \hat{\phi}. \tag{3.17}$$

This is analogous to the decomposition (2.7) in the sharp momentum cut-off scheme. We then treat $\hat{\phi}$ as the field to integrate over in the functional integral whilst treating φ as a fixed background field.

Since φ is slowly varying we can expand in powers of derivatives of φ and then, as in Chap. 2, in order to calculate the effective potential we can assume that φ is constant. The effective potential in $d = 4$ to one loop is then

$$V_{\text{eff}}(\varphi) = V(\varphi) + \frac{1}{2}\mu^{4-d} \int \frac{d^d p}{(2\pi)^d} \log\left(p^2 + V''(\varphi)\right) \tag{3.18}$$

$$= V(\varphi) - \frac{V''(\varphi)^2}{32\pi^2\varepsilon} + \frac{V''(\varphi)^2}{64\pi^2}\left[-\frac{3}{2} + \gamma_E + \log\frac{V''(\varphi)}{4\pi\mu^2}\right].$$

Subtracting the divergence with a counter-term in the $\overline{\text{MS}}$ scheme gives

$$V_{\text{eff}}(\varphi) = V(\varphi) + \frac{V''(\varphi)^2}{64\pi^2}\left[-\frac{3}{2} + \log\frac{V''(\varphi)}{\mu^2}\right]. \tag{3.19}$$

With a potential $V(\phi) = \frac{1}{2}m^2\phi^2 + \frac{1}{4!}\lambda\phi^4$, notice that the counter-term is proportional to

$$V''(\varphi)^2 = \left(m^2 + \frac{1}{2}\lambda\varphi^2\right)^2, \tag{3.20}$$

which has terms proportional to φ^2 and φ^4 and so takes the same form as the original potential. In other words, we only need to add counter-terms for m^2 and λ. With a little more work one finds that the counter-terms are precisely (3.11) and (3.14).

The RG flow of the dimensionless couplings $g_2 = m^2/\mu^2$ and $g_4 = \lambda$ can be deduced, as before, by requiring that the effective potential satisfies an RG equation. For instance, in the form (1.4),

$$V_{\text{eff}}(Z(\mu)^{1/2}\varphi; \mu, g_i(\mu)) = V_{\text{eff}}(Z(\mu')^{1/2}\varphi; \mu', g_i(\mu')), \tag{3.21}$$

where here μ is not the Wilsonian cut off, but is a variable that plays an analogous rôle, the mass scale introduced in dimensional regularization.

To one-loop order there is no wave-function renormalization and, using the above, we find

$$\mu \frac{dg_2}{d\mu} = -2g_2 + ag_2 g_4,$$

$$\mu \frac{dg_4}{d\mu} = 3ag_4^2, \tag{3.22}$$

where $a = 1/16\pi^2$. It is interesting to compare (3.22) with the beta functions calculated in the momentum cut-off scheme (2.22) (with $d = 4$ and with the irrelevant couplings set to 0). The first difference is that with the latter scheme the beta functions are exact at the one-loop level, whereas (3.22) receives contributions to all loop orders. Secondly, the momentum cut-off scheme displays manifest decoupling when $m \gg \mu$ ($g_2 \gg 1$), whereas, on the contrary, dimensional regularization with the $\overline{\text{MS}}$ subtraction scheme does not. In the next chapter we will explore this in more detail and explain how decoupling must be implemented by hand in the $\overline{\text{MS}}$ scheme. What these two schemes illustrate is that the actual RG flows depend in detail on the chosen scheme. However, it is a central feature of the theory that the "topological" properties of the flows, meaning existence of fixed points, or whether a coupling is relevant or irrelevant, is scheme independent. For example, if one repeats the analysis above in dimensions $d < 4$ rather than $d = 4$, then one can demonstrate the existence of the Wilson-Fisher fixed point.

3.2 Triviality

Our analysis of scalar field theories leads to the following important conclusion for space-time dimension $d = 4$. Since there is only a single fixed point, namely the non-interacting Gaussian fixed point, the only possible continuum theory is that of a free massive or massless particle. The lack of an interacting continuum scalar theory in 4 dimensions is known as *triviality*.

This conclusion seems at odds with perturbation theory where it seems that we can define an interacting theory with a finite renormalized coupling. However, something goes terribly wrong with perturbation theory caused by the need to keep the renormalized coupling finite whilst sending the cut off $\mu' \to \infty$. We can see what goes wrong by looking at the flow of the coupling (3.2): with μ fixed then as μ' increases there is a singularity at

$$\mu' = \mu e^{16\pi^2/3g_4(\mu)}, \tag{3.23}$$

where the bare coupling $g_4(\mu')$ diverges. This is known as a *Landau pole* (this terminology is properly associated with QED which, as we shall see in Chap. 4, suffers the same fate). Of course this is exactly what we expected: the coupling

$g_4(\mu)$ is irrelevant at the Gaussian fixed point and so when we flow backward along the RG trajectory towards the UV, the path veers away from the fixed point. Now we see that, within the one-loop approximation, the flow actually diverges and becomes infinite at a finite value of μ'. So within the level of approximation we are working, the conclusion is that in the absence of a non-trivial fixed point, scalar QFT is not a renormalizable theory in $d = 4$, even though it appears to be renormalizable order-by-order in perturbation theory. It is possible to define an interacting effective theory but this theory will have an explicit cut off and so will only make sense for momenta $p < \mu'$. This conclusion applies to the Higgs sector of the standard model, and so this sector of the standard model is not truly a renormalizable theory and predicts its own demise at high enough energies when the momenta reach the cut off scale μ' as determined by the coupling $g_4(\mu)$ at the low momentum scale μ.

Summary

We now sketch how to use perturbation theory (when valid) and the background field method to calculate the RG flows of the relevant couplings.

(1) Write down a Lagrangian $\mathscr{L}_r(\phi)$ with all the "relevant" couplings g_i^r.
(2) Split $\phi = \varphi + \hat{\phi}$ and calculate loop diagrams with external φ and internal $\hat{\phi}$ (more precisely only include diagrams which are *one particle irreducible* 1PI) and add counter-terms along the way to cancel divergences. It is important that the counter-terms have the same form as $\mathscr{L}_r(\varphi)$ (if not then the definition of the $\mathscr{L}_r(\varphi)$ needs to be suitably generalized).
(3) The analogue of the Wilsonian effective action in the special subspace of theories parametrized by the relevant coupling is then

$$S[\varphi; \mu, g_i^r] = \int d^d x \left(\mathscr{L}_r(\varphi) + \mathscr{L}_{ct}(\varphi) + \mathscr{L}_{1PI}(\varphi) \right). \qquad (3.24)$$

Note that this action will include all the irrelevant couplings as well; however, these will depend on the relevant couplings.
(4) Extract the RG flows of the relevant couplings in $\mathscr{L}_r(\varphi)$ by imposing the RG equation (1.4).

3.3 RG Improvement

In this problem we consider the issue of RG improvement in the context of the effective potential of a ϕ^4 theory in $d = 4$. The one-loop effective potential in the $\overline{\text{MS}}$ scheme is given in (3.19). Taking $V = \frac{1}{2}m^2\phi^2 + \frac{1}{4!}\lambda\phi^4$, we have

$$V_{\text{eff}}(\varphi) = \frac{1}{2}m^2\varphi^2 + \frac{1}{4!}\lambda\varphi^4 + \frac{(m^2 + \frac{1}{2}\lambda\varphi^2)^2}{64\pi^2}\left[-\frac{3}{2} + \log\frac{m^2 + \frac{1}{2}\lambda\varphi^2}{\mu^2}\right]. \quad (3.25)$$

Now the pertinent question is: what are couplings m and λ in the above? The guiding principle is that V_{eff} should satisfy the RG equation (3.21), and so it appears that the couplings m and λ should run with μ in such a way that V_{eff} is independent of μ. The running couplings $m(\mu)^2 = \mu^2 g_2(\mu)$ and $\lambda(\mu) = g_4(\mu)$ are the solutions of the beta function equations (3.22) and to one-loop order there is no wave-function renormalization.

Since V_{eff} is supposed to be μ-independent it should not matter what value we take for μ. However, the choice of μ turns out to be important as a result of the inevitable short-comings of perturbation theory. If we could sum up the whole of the perturbative series then indeed it would not matter what we take for μ; however, as soon as we work to a given order in perturbation theory, we have to perform a truncation in powers of the running coupling and then the truncated effective potential does depend on μ.

In order to simplify the remaining discussion, let us specialize to the massless case and point out an interesting—but in retrospect, incorrect—conclusion. Setting $m = 0$, we have

$$V_{\text{eff}}(\varphi) = \frac{1}{4!}\lambda\varphi^4 + \frac{\lambda^2\varphi^4}{256\pi^2}\left[-\frac{3}{2} + \log\frac{\lambda\varphi^2}{2\mu^2}\right]. \quad (3.26)$$

This apparently has a minimum away from the origin at a value of φ for which

$$\varphi^2 = \frac{2\mu^2}{\lambda}\exp\left[1 - \frac{32\pi^2}{3\lambda}\right]. \quad (3.27)$$

So it seems that the one-loop correction leads to an effective potential whose minimum is pushed away from the origin and a VEV for the scalar field will be generated. This might have interesting applications for spontaneous symmetry breaking. However, our suspicions should be raised because the value of φ above is clearly non-perturbative in the coupling and this has as a consequence that, with this value of the field, the tree-level and one-loop contributions to the effective potential are of the same order of magnitude. Clearly the perturbative expansion is being violated and we cannot trust the conclusion.

The problem is that the log in (3.26) becomes large as $\varphi \to 0$ invalidating the perturbative expansion around the tree-level minimum $\varphi = 0$. However we can repair the small φ behaviour of $V_{\text{eff}}(\varphi)$ by recognizing that $V_{\text{eff}}(\varphi) = V_{\text{eff}}(\varphi; \mu, \lambda(\mu))$ and then making an appropriate choice for the RG scale μ. Since there is only one additional scale in the problem, namely φ, the natural choice is to take $\mu = \varphi$, so that[2]

$$V_{\text{eff}}(\varphi) = V_{\text{eff}}(\varphi; \varphi, \lambda(\varphi))$$

$$= \frac{1}{4!}\lambda(\varphi)\varphi^4 + \frac{\lambda(\varphi)^2\varphi^4}{256\pi^2}\left[-\frac{3}{2} + \log\frac{\lambda(\varphi)}{2}\right]. \tag{3.28}$$

Remember that to all orders in perturbation theory this has not changed the effective potential, however, once we truncate to one loop, the functional behaviour of this effective potential is completely different from (3.26). In particular, in the limit $\varphi \to 0$ the running coupling, which to one-loop order is the solution to the equation

$$\mu\frac{d\lambda}{d\mu} = \frac{3}{16\pi^2}\lambda^2, \tag{3.29}$$

is

$$\frac{1}{\lambda(\varphi)} = C - \frac{3}{16\pi^2}\log\varphi \tag{3.30}$$

and therefore goes to 0 as $\varphi \to 0$. So as $\varphi \to 0$, perturbation theory gets better and better. The "RG improved" effective potential now has a minimum at $\varphi = 0$ and therefore there is no spontaneous symmetry breaking. The moral of this story is that for the success of perturbation theory, it should be performed in the coupling which runs with the energy scale of the problem at hand and is only valid in a region where it is small.

Coleman and Weinberg (1973) realized that a consistent model of spontaneous symmetry driven breaking by quantum loop effects can be constructed in a theory with more than one coupling constant. For instance, if the scalar field is coupled to the electromagnetic field there are two couplings: λ, as above, and also the electric charge e of the scalar field. The scalar field must necessarily be complex. The net result is that the effective potential now receives a contribution from a photon loop leading to an expression similar to (3.26), but with λ in the second term replaced by e^2 and an additional factor of 3 coming from the extra degrees-of-freedom of the photon:

$$V_{\text{eff}}(\varphi) = \frac{1}{4!}\lambda\varphi^4 + \frac{3e^4\varphi^4}{256\pi^2}\left[-\frac{3}{2} + \log\frac{e^2\varphi^2}{2\mu^2}\right]. \tag{3.31}$$

The contribution from the photon loop dominates the original scalar loop and the latter is ignored. Now the minimum of (3.31) lies at

$$\varphi^2 = \frac{2\mu^2}{e^2}\exp\left[1 - \frac{32\pi^2\lambda}{9e^4}\right]. \tag{3.32}$$

---

[2] At the one-loop level we can ignore wave-function renormalization. In addition, note that any other choice $\mu = a\varphi$ for a constant $a \neq 1$ is equivalent to $a = 1$ with a re-definition of the mass and coupling.

In this case, just as previously, the existence of this minimum involves balancing a tree-level and one-loop contribution, but now for two different perturbative expansions: in λ and e^2, respectively. Consequently, unlike the previous situation the conclusion that a VEV develops can now be trusted.

Bibliographical Notes

The RG first appeared in the context of perturbation theory and the ultra-violet divergences that arise. This aspect of the RG is described, for instance, in the textbooks by Collins (1984) and Peskin and Schroeder (1995). The latter book also describes the process of RG improvement.

References

Coleman, S.R., Weinberg, E.J.: Radiative corrections as the origin of spontaneous symmetry breaking. Phys. Rev. D **7**, 1888 (1973)

Collins, J.C: Renormalization. An Introduction to Renormalization, the Renormalization Group, and the Operator Product Expansion, p. 380. University Press, Cambridge (1984)

Peskin, M.E., Schroeder, D.V.: An Introduction to Quantum Field Theory. Addison-Wesley, Reading (1995)

Chapter 4
Gauge Theories and Running Couplings

In this chapter, we turn our attention to the RG properties of gauge theories including QED along with the strong and weak interactions.

4.1 Quantum Electro-Dynamics

We begin with the simplest gauge theory QED. For simplicity, we also start with the case of "scalar QED" where the rôle of the electron is played by a charged scalar field. The scalar field ϕ must necessarily be complex, and for simplicity we take a Lagrangian without any self-interactions for ϕ:

$$\mathscr{L} = -\frac{1}{4} F_{\mu\nu} F^{\mu\nu} - |D_\mu \phi|^2 - m^2 |\phi|^2, \tag{4.1}$$

where the covariant derivative $D_\mu \phi = \partial_\mu \phi + ieA_\mu \phi$ and $F_{\mu\nu} = \partial_\mu A_\nu - \partial_\nu A_\mu$. Here e is the electric charge of the field ϕ. The theory is invariant under gauge transformations:

$$\phi(x) \to e^{ie\alpha(x)} \phi(x), \qquad A_\mu(x) \to A_\mu(x) - \partial_\mu \alpha(x). \tag{4.2}$$

We want to focus on the RG flow of the electric charge e. In order to do this it is actually more convenient to re-scale $A_\mu \to A_\mu/e$ so that the coupling now appears in front of the photon's kinetic term and is interpreted as a gauge coupling constant rather than a charge

$$\mathscr{L} = -\frac{1}{4e^2} F_{\mu\nu} F^{\mu\nu} - |(\partial_\mu + iA_\mu)\phi|^2 - m^2 |\phi|^2. \tag{4.3}$$

The reason why this is a convenient choice in the context of the RG is that under the flow, the structure of the covariant derivative must remain intact otherwise gauge

T. J. Hollowood, *Renormalization Group and Fixed Points*, SpringerBriefs
in Physics, DOI: 10.1007/978-3-642-36312-2_4, © The Author(s) 2013

invariance would not be respected. This means that A_μ with the Lagrangian as above must not undergo any wave-function renormalization, rather we should interpret any renormalization of the gauge kinetic term as a renormalization of the electric charge, or gauge coupling constant, e.[1]

In order to find the RG flow of the gauge coupling e we use the background field method, treating A_μ as the background field and ϕ as the fluctuating field that is to be integrated out of the functional integral. The interaction terms are

$$\mathscr{L}_{\text{int}} = -iA_\mu \left(\phi\partial^\mu\phi^* - \phi^*\partial^\mu\phi\right) - A_\mu A^\mu |\phi|^2. \tag{4.4}$$

At the one-loop level there are 2 relevant diagrams, shown below, that contribute to the photon's kinetic term:

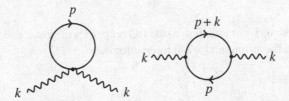

The first one cannot depend on the photon's momentum k since this does not flow around the loop. Its rôle is to cancel the k-independent contribution of the second diagram. In dimensional regularization the second diagram yields a contribution proportional to

$$\mu^{4-d} \int \frac{d^d p}{(2\pi)^d} \cdot \frac{(2p+k)_\mu (2p+k)_\nu}{\left(p^2 + m^2\right)\left((p+k)^2 + m^2\right)}. \tag{4.5}$$

The standard trick for dealing with a product of propagators is to introduce a Feynman parameter, which in this case involves using the identity

$$\frac{1}{p^2 + m^2} \cdot \frac{1}{(p+k)^2 + m^2} = \int_0^1 dx \, \frac{1}{\left((1-x)(p^2 + m^2) + x((p+k)^2 + m^2)\right)^2} \tag{4.6}$$

and then one shifts $p \to p - xk$. Having done this (4.5) becomes[2]

$$\mu^{4-d} \int_0^1 dx \int \frac{d^d p}{(2\pi)^d} \cdot \frac{(2p + (1-2x)k)_\mu (2p + (1-2x)k)_\nu}{\left(p^2 + x(1-x)k^2 + m^2\right)^2}$$

---

[1] Of course this will only be true if the cut-off scheme itself does not break gauge invariance. For instance the sharp momentum cut off used in Chap. 2 is not consistent with gauge invariance because gauge transformations inevitably mix up modes of different frequency, i.e. they do not preserve the split of modes in (2.7).

[2] The momentum integrals below are standard and can be found, for example, in Appendix A of the text-book of Peskin and Schroeder (1995).

$$= \mu^{4-d} \frac{\Gamma(1-\frac{d}{2})}{(4\pi)^{d/2}} \int_0^1 dx \, \Delta^{d/2-2}\Big[(1-\tfrac{d}{2})(1-2x)^2 k_\mu k_\nu + 2\Delta\eta_{\mu\nu}\Big], \qquad (4.7)$$

where $\Delta = x(1-x)k^2 + m^2$. The second term in the bracket can be re-written using the identity

$$2\int_0^1 dx \, \Delta^{d/2-1} = k^2(\tfrac{d}{2}-1)\int_0^1 dx \, (1-2x)^2 \Delta^{d/2-2} + 2m^{d-2}. \qquad (4.8)$$

The last term here cancels the contribution

$$\mu^{4-d} \int \frac{d^d p}{(2\pi)^d} \frac{\eta_{\mu\nu}}{p^2+m^2} = \mu^{4-d} \frac{\Gamma(1-\frac{d}{2})}{(4\pi)^{d/2}} m^{d-2}\eta_{\mu\nu} \qquad (4.9)$$

from the first diagram above. After this cancellation, what remains is

$$\frac{\mu^{4-d}}{(4\pi)^{d/2}} \big(k_\mu k_\nu - k^2\eta_{\mu\nu}\big)\Gamma(2-\tfrac{d}{2})\int_0^1 dx \, (1-2x)^2\big(x(1-x)k^2+m^2\big)^{d/2-2}. \quad (4.10)$$

The implications of this result for the effective action become clearer when we expand in powers of k. What results are terms of the form

$$k^{2n}\big(k_\mu k_\nu - k^2\eta_{\mu\nu}\big) \qquad (4.11)$$

in momentum space which correspond to terms schematically of the form $\partial^n F_{\mu\nu}\partial^n F^{\mu\nu}$ in the effective action. This is because in momentum space

$$F_{\mu\nu}(-k)F^{\mu\nu}(k) = -\big(ik_\mu A_\nu(-k) - ik_\nu A_\mu(-k)\big)\big(ik^\mu A^\nu(k) - ik^\nu A^\mu(k)\big)$$
$$= -2\big(k_\mu k_\nu - k^2\eta_{\mu\nu}\big)A^\mu(-k)A^\nu(k). \qquad (4.12)$$

Note that since the field strength $F_{\mu\nu}$ itself is gauge invariant, the effective action we have derived is gauge invariant. Taking the limit $\varepsilon = 4 - d \to 0$ of (4.10) leads to

$$\int_0^1 dx \, (1-2x)^2\Big[\frac{1}{8\pi^2\varepsilon} - \frac{\gamma_E}{16\pi^2} - \frac{1}{16\pi^2}\log\frac{x(1-x)k^2+m^2}{4\pi\mu^2}\Big] \qquad (4.13)$$

times $k_\mu k_\nu - k^2\eta_{\mu\nu}$. In the $\overline{\text{MS}}$ scheme, the divergence is cancelled by a counter-term

$$\mathscr{L}_{\text{ct}} = -\frac{1}{12}\Big[-\frac{1}{8\pi^2\varepsilon} + \frac{\gamma_E - \log 4\pi}{16\pi^2}\Big]F_{\mu\nu}F^{\mu\nu}, \qquad (4.14)$$

where we have used

$$\int_0^1 (1-2x)^2 = \frac{1}{3}. \qquad (4.15)$$

Being careful with the overall normalization, we have derived the terms in the one-loop effective action for the photon, which are quadratic in A_μ, in the $\overline{\text{MS}}$ scheme:

$$S_{\text{eff}}[A_\nu] = -\frac{1}{4}\int\frac{d^4k}{(2\pi)^4}\left[\frac{1}{e^2}\right.$$
$$\left. -\frac{1}{16\pi^2}\int_0^1 dx\,(1-2x)^2\log\frac{x(1-x)k^2+m^2}{\mu^2}\right]F_{\mu\nu}(-k)F^{\mu\nu}(k).$$
$$(4.16)$$

The one-loop beta function in the $\overline{\text{MS}}$ scheme can then be deduced by demanding that S_{eff} satisfies an RG equation of the form (1.4), but with no wave-function renormalization in this case:

$$S_{\text{eff}}[A_\nu;\mu,e(\mu)] = S_{\text{eff}}[A_\nu;\mu',e(\mu')]. \qquad (4.17)$$

This instantly yields the beta function of the gauge coupling, or electric charge, in the $\overline{\text{MS}}$ scheme:

$$\mu\frac{de}{d\mu} = \frac{e^3}{16\pi^2}\int_0^1 dx\,(1-2x)^2 = \frac{e^3}{48\pi^2}. \qquad (4.18)$$

It is interesting to repeat the calculation for QED with more conventional spinor matter. The electron and positron correspond to a Dirac fermion; however, we will work in terms of Weyl fermions which couple to the photon via a kinetic term with a covariant derivative[3]

$$\mathscr{L}_{\text{kin}} = i\bar{\psi}^{\dot\alpha}\bar{\sigma}^\mu_{\ \dot\alpha\alpha}D_\mu\psi^\alpha. \qquad (4.19)$$

In this case the fermion contributes to the vacuum polarization via the single diagram

If one follows the derivation, the only difference from the scalar case in the final result amounts to the replacement of $(1-2x)^2 \to 4x(1-x)$ in (4.16). In general, if we have a set of complex scalars and Weyl fermions with integer charges q_i, meaning that covariant derivatives involve $D_\mu = \partial_\mu + ieq_i A_\mu$, or $\partial_\mu + iq_i A_\mu$ after the rescaling $A_\mu \to A_\mu/e$, then the one-loop beta function in the $\overline{\text{MS}}$ scheme is

$$\mu\frac{de}{d\mu} = \left(\frac{1}{48\pi^2}\sum_{\text{scalars}}q_i^2 + \frac{1}{24\pi^2}\sum_{\text{fermions}}q_i^2\right)e^3. \qquad (4.20)$$

---

[3] A Weyl fermion ψ_α, $\alpha = 1,2$ is a 2-component complex Grassmann quantity. In Minkowski space its conjugate is denoted $\bar{\psi}^{\dot\alpha} = (\psi_\alpha)^\dagger$, $\dot\alpha = 1,2$. A Weyl fermion can also be represented in 4-component Dirac fermion language as a Majorana fermion $\Psi = (\psi_\alpha\ \ \bar{\psi}^{\dot\alpha})$. A Dirac fermion, like the electron-positron field, is then made up of two Weyl fermions $\Psi = (\psi_\alpha\ \ \bar{\lambda}^{\dot\alpha})$. In this 2-component notation the σ matrices can be taken to be $\sigma^\mu_{\ \alpha\dot\alpha} = (\mathbf{1},\sigma^i)$, $\bar{\sigma}^{\mu\dot\alpha\alpha} = (\mathbf{1},-\sigma^i)$.

Solving (4.20) for the running coupling yields

$$\frac{1}{e(\mu)^2} = C - \frac{b}{24\pi^2} \log \mu \quad \text{or} \quad e(\mu)^2 = \frac{24\pi^2}{b \log(\Lambda/\mu)}, \tag{4.21}$$

where $b = \sum_{\text{scalars}} q_i^2 + 2 \sum_{\text{fermions}} q_i^2 > 0$. Notice that e, like the coupling g_4 in scalar field theory, is irrelevant at the Gaussian fixed point. So $e(\mu)$ becomes smaller in the IR but has a Landau pole in the UV—at least within the one-loop approximation. Hence QED, like scalar QFT, does not seem to have a continuum limit in $d = 4$. Of course this does not rule out a more complicated "UV completion", an issue that we will return to later.

4.2 Decoupling in $\overline{\text{MS}}$

There is an important subtlety with regularization schemes like $\overline{\text{MS}}$ that do not involve setting renormalization conditions at a particular momentum, or energy, scale. If we compare the beta functions in (2.22) and (3.22) then the former exhibit manifest decoupling, meaning that if we start with some cut off $\mu > m$ and flow down towards the IR then when $\mu < m$ the contribution to the RG flow from the particle is suppressed. This makes perfect physical sense: a particle of mass m only has effects when the momentum scale in question is greater than m. However, in the $\overline{\text{MS}}$ scheme (3.22) (or in the case of QED (4.18)) there does not appear to be any decoupling at all since the coupling is still running when $\mu \ll m$. While there is nothing wrong with this in principle, it is not what one expects of the physical coupling. In fact if one looks at the perturbation expansion in powers of $e(\mu) \to 0$ of (4.16), then the logarithmic behaviour of the running coupling as $\mu \to 0$ is actually compensated by the factor in the second term involving $\log \mu \to -\infty$.

These facts simply illustrate the fact that the couplings in the Wilsonian action need not be the physically observable couplings of the theory. The situation suggests that we can introduce a more physically motivated coupling in a momentum-independent regularization scheme like $\overline{\text{MS}}$ by decoupling the massive particle by hand when the RG scale μ crosses the threshold at $\mu = m$. The recipe is as follows: two theories are written down, one including the field valid for $\mu \geq m$ and one without the field for $\mu \leq m$. At $\mu = m$ physical quantities in the two theories are matched. For example, in scalar QED, for $\mu \geq m$ the coupling $e(\mu)$ runs as in (4.18). At $\mu = m$ the electron is removed from the theory to leave the photon by itself. This is a non-interacting theory in which the couplings does not run. After matching the couplings at $\mu = m$, the coupling in the low energy theory $\mu < m$ is therefore frozen at the value $e(m)$. The nasty $\log \mu$ factors disappear from perturbation theory to be left with a nice power series in $e(m)$ that is a useful series if $e(m)$ is small enough, which in QED itself is the case since $\frac{e(m)^2}{4\pi} = \frac{1}{137}$. For energy scales $<m$, the effective theory is therefore (4.16) with $\mu \to m$ and $e \equiv e(m)$. At low energies we can expand the

logarithm in powers of k^2 and then Fourier transform back to real space to give terms quadratic in the field strength with an arbitrary number of derivatives. In the series the first two terms are

$$S_{\text{eff}}[A_\nu] = -\frac{1}{4} \int d^4x \left[\frac{1}{e^2} F_{\mu\nu} F^{\mu\nu} - \frac{1}{480\pi^2 m^2} \partial_\sigma F_{\mu\nu} \partial^\sigma F^{\mu\nu} + \cdots \right]. \quad (4.22)$$

The effective action (4.16) also illustrates what goes wrong as we try to push the energy scale beyond the region of validity of the effective theory. A physical photon satisfies the on-shell condition $k^2 = 0$, however, if we consider an off-shell photon with $k^2 < 0$, then when $k^2 = -4m^2$ the integral over x hits the branch cut of the logarithm at the point $x = \frac{1}{2}$. As k^2 decreases beyond $-4m^2$ the integral over x picks up an imaginary part. It appears that the effective action has become imaginary, and this signals the fact that the effective theory by itself is non-unitary. In fact the imaginary part has a very obvious physical interpretation as being due to the physical process involving the decay of the virtual photon into pairs of scalar particles $\gamma \rightarrow \phi\phi$. Indeed the kinematic threshold for this process is precisely at $k^2 = -4m^2$. So the effective description which includes only the photon has broken down and if we want to regain a unitarity description with a real action, then we need to re-introduce the scalar particle back into the effective action.

In this more general philosophy that we use with the $\overline{\text{MS}}$ scheme, as one flows into the IR, particles are decoupled and removed from the theory by hand as the RG scale μ passes the appropriate mass thresholds. On the contrary, in a mass-dependent RG scheme, decoupling is automatic. For example, if instead of performing minimal subtraction we instead fix the value of the effective coupling in (4.16) at the space-like momentum scale $k^2 = \tilde{\mu}^2$. This requires a counter-term

$$\mathscr{L}_{\text{ct}} = -\frac{1}{4} \int_0^1 dx\, (1 - 2x)^2 \left[\frac{1}{8\pi^2 \varepsilon} - \frac{1}{16\pi^2} \log \frac{x(1-x)\tilde{\mu}^2 + m^2}{\mu^2} \right] F_{\mu\nu} F^{\mu\nu}. \quad (4.23)$$

We now think of $\tilde{\mu}$ as the RG scale and the beta function in this scheme takes the form

$$\tilde{\mu} \frac{de}{d\tilde{\mu}} = \frac{e^3}{16\pi^2} \int_0^1 dx\, (1 - 2x)^2 \frac{\tilde{\mu}^2 x(1-x)}{m^2 + \tilde{\mu}^2 x(1-x)}, \quad (4.24)$$

which, unlike (4.18), does indeed have suppressed RG flow when $\tilde{\mu} \ll m$.

4.3 Non-Abelian Gauge Theories

Now we turn to the generalization of the RG flow in gauge theories to the non-abelian case. The Lagrangian for a non-abelian gauge field coupled to a massive scalar field transforming in a representation r of the gauge group G, has the form

$$\mathscr{L} = -\frac{1}{4g^2} F_{\mu\nu}^a F^{a\,\mu\nu} - \left| D_\mu \phi_i \right|^2 - m^2 |\phi_i|^2, \quad (4.25)$$

where the non-abelian field strength is

$$F_{\mu\nu}^a = \partial_\mu A_\nu^a - \partial_\nu A_\mu^a + f^{abc} A_\mu^b A_\nu^c \tag{4.26}$$

and the covariant derivative is

$$D_\mu \phi_i = \partial_\mu \phi_i + i A_\mu^a T_{ij}^a \phi_j, \tag{4.27}$$

where f^{abc} are the structure constants of the gauge group G and T_{ij}^a are the generators of a representation r of G, so that as matrices

$$[T^a, T^b] = i f^{abc} T^c. \tag{4.28}$$

The RG flow of the gauge coupling g can be determined in a similar way to QED but the details are a good deal more complicated. Integrating out the matter field produces a contribution which is exactly (4.18) but multiplied by the group theory factor $C(r)$ where $C(r)\delta_{ab} = \text{Tr}(T^a T^b)$. For example, for the gauge group SU(N), for the adjoint representation $r = G$, $C(G) = N$, and for the N-dimensional fundamental, or defining, representation $C(\text{fund}) = \frac{1}{2}$. However, the new feature is that there are contributions from the gauge field itself because of its self-interactions. The relevant diagrams are[4]

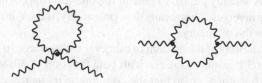

The resulting expression for the one-loop beta function, including the contributions from a series of scalar fields, as above, and, in addition, Weyl fermions in representations r_f, all with a mass $m < \mu$, is

$$\mu \frac{dg}{d\mu} = -\frac{g^3}{(4\pi)^2} \left(\frac{11}{3} C(G) - \frac{1}{3} \sum_{\text{scalars } f} C(r_f) - \frac{2}{3} \sum_{\text{fermions } f} C(r_f) \right). \tag{4.29}$$

For example, for gauge group $G = $ SU(N), N_s complex scalars and N_f Weyl fermions in the fundamental N-dimensional representation plus the anti-fundamental $\bar{N}$ (equivalently N_f fundamental Dirac fermions),

$$\mu \frac{dg}{d\mu} = -\frac{g^3}{(4\pi)^2} \left(\frac{11}{3} N - \frac{1}{3} N_s - \frac{2}{3} N_f \right). \tag{4.30}$$

---

[4] Depending on how the gauge is fixed, there will usually be a contribution from the gauge-fixing ghost fields. The complete calculation can be found in the text-book of Peskin and Schroeder (1995) from p. 533 onwards.

On the other hand, if all fields are in the adjoint representation

$$\mu \frac{dg}{d\mu} = -\frac{Ng^3}{(4\pi)^2}\left(\frac{11}{3} - \frac{1}{3}N_s - \frac{2}{3}N_f\right). \tag{4.31}$$

What is remarkable is that the contribution from the self interactions of the gauge field have an opposite sign from the matter fields and this has important implications for the running of the gauge coupling. If we write the one-loop beta function as $\mu\, dg/d\mu = -\beta_1 g^3/(4\pi)^2$ then the resulting behaviour of the RG flow depends crucially on the sign of β_1. Solving the one-loop beta function equation gives

$$g(\mu)^2 = \frac{8\pi^2}{C + \beta_1 \log\mu} = \frac{8\pi^2}{\beta_1 \log(\mu/\Lambda)}, \tag{4.32}$$

where we have written the integration constant as a parameter Λ with unit mass dimension. Once again we have an example of *dimensional transmutation* where the dimensionless gauge coupling in the original action has turned into the freedom to specify Λ, a parameter with a non-trivial mass dimension. When there is sufficient matter so that $\beta_1 < 0$, the situation will be as in QED. In this case, we must have $\mu < \Lambda$, $g(\mu)$ will decrease along the flow and the theory is IR free. The coupling is marginally irrelevant at the Gaussian fixed point $g = 0$ and as we flow backwards into the UV there is actually a Landau pole in the one-loop approximation at $\mu = \Lambda$ where the coupling diverges. This means that, potentially, there is no continuum limit for QED just as for the ϕ^4 theory.

On the contrary, as long as the amount of matter is not too large so that $\beta_1 > 0$ (like in QCD itself with 3 flavours of quarks with beta function (4.30) with $N_s = 0$ and $N_f = 3$) we must have $\mu > \Lambda$, so that the coupling $g(\mu)$ increases along the flow. In this case the coupling is marginally relevant at the Gaussian fixed point, which lies in the UV, and a non-trivial interacting continuum limit of the theory exists since there is a renormalized trajectory that flows out of the Gaussian fixed point. The UV limit of the theory is consequently non-interacting, a situation known as asymptotic freedom. In this case, Λ signals the scale at which the coupling becomes large and perturbation theory breaks down. In QCD this is the scale at which the phenomenon of confinement sets in. It is important to realize that in a theory like QCD, Λ is a physically measurable energy scale. Below we show the behaviour of the RG flow of the gauge coupling in the two scenarios:

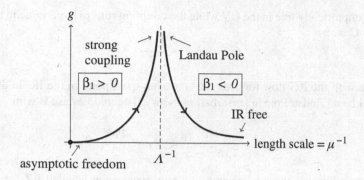

4.4 Banks-Zaks Fixed Points

It is interesting to investigate the RG flow of the gauge coupling of non-abelian gauge theories beyond the one-loop approximation. The beta-function of SU(N) gauge theory with N_f flavours of quarks in the N representation and N_f in the $\bar{N}$ representation, is, to the 2-loop level,

$$\mu \frac{dg}{d\mu} = -\frac{\beta_1}{(4\pi)^2}g^3 - \frac{\beta_2}{(4\pi)^4}g^5 \qquad (4.33)$$

with

$$\beta_1 = \frac{11N}{3} - \frac{2N_f}{3}, \qquad \beta_2 = \frac{34N^2}{3} - \frac{10NN_f}{3} - \frac{N_f(N^2-1)}{N}. \qquad (4.34)$$

In the diagram we plot the beta function and direction of RG flows showing how the behaviour changes as N_f is varied for fixed N.

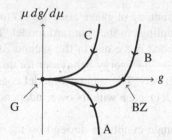

The diagram shows the Gaussian fixed point **G**. The case **A** has

$$\mathbf{A}: \quad N_f < \frac{34N^3}{13N^2 - 3}$$

and is asymptotically free in the UV while the coupling runs to large value in the IR. The case **C** with

$$\mathbf{C}: \quad N_f > \frac{11N}{2}$$

is IR free with the RG flow towards the Gaussian fixed point in the IR. In the UV, there will be a Landau Pole in perturbation theory. The middle case **B** with

$$\mathbf{B}: \quad \frac{34N^3}{13N^2 - 3} < N_f < \frac{11N}{2}$$

is interesting because there is apparently a new fixed point labelled **BZ** for Banks and Zaks (1982). In this case, there should be a continuum massless theory whose physics is controlled by the **BZ** fixed point in the IR and by the Gaussian fixed point in the UV. The allowed region for N_f above is known as the *conformal window*. The question is whether we can be certain that the new fixed point actually exists, given the limited applicability of perturbation theory. The fixed-point value of the coupling is

$$g_*^2 = (4\pi)^2 \frac{11N^2 - 2NN_f}{34N^3 - 13NN_f + 3N_f}. \tag{4.35}$$

This can be small near the top of the conformal window at $N_f = 11N/2$; for instance, for QCD itself $N = 3$, the top of the conformal window is at $N_f = 16$, which gives $g_*^2 \sim 0.52$. For larger N, g_*^2 near the top of the conformal window scales like N^{-1} for large N and perturbation theory becomes more reliable. Correspondingly, near the bottom of the conformal window the fixed-point coupling g_*^2 is large and perturbation theory cannot be trusted.

4.5 The Standard Model and Grand Unification

One application of the RG running of gauge couplings is to the high and low energy behaviour of the gauge couplings in the standard model. The appropriate interpretation of the RG that is needed is the one in the second RG equation, (1.3), which allows us to trade the high or low energy behaviour for an RG flow of the coupling into the UV and IR, respectively. The standard model consists of 3 separate gauge groups SU(3) × SU(2) × U(1), each with its own independent coupling constant g_i and its own RG flow.

The RG flows of the gauge couplings depend on the spectrum of fields in the standard model and how they transform under the three gauge groups. The running is most conveniently written in terms of $\alpha_i = g_i^2/4\pi$, with $i = 1, 2, 3$ for U(1), SU(2) and SU(3), respectively. At one-loop order, solving the beta function equation gives

$$\frac{1}{\alpha_i(\mu)} = \frac{1}{\alpha_i(\mu')} + \frac{\beta_1^{(i)}}{2\pi} \log \frac{\mu}{\mu'}. \tag{4.36}$$

The one-loop coefficients $\beta_1^{(i)}$ are determined by the number of families N_{fam} of matter fields and the number of Higgs doublets N_{Higgs}:

$$\beta_1^{(1)} = -\frac{4}{3}N_{\text{fam}} - \frac{1}{10}N_{\text{Higgs}},$$

$$\beta_1^{(2)} = \frac{22}{3} - \frac{4}{3}N_{\text{fam}} - \frac{1}{6}N_{\text{Higgs}}, \qquad (4.37)$$

$$\beta_1^{(3)} = 11 - \frac{4}{3}N_{\text{fam}}.$$

In the standard model there are three families $N_{\text{fam}} = 3$ and one Higgs doublet $N_{\text{Higgs}} = 1$. From this we see that the QCD coupling g_3 is asymptotically free and is therefore weak in the UV and runs to strong coupling in the IR. The SU(2) coupling is also asymptotically free; however, its running is also affected by the Higgs mechanism which leads to spontaneous symmetry breaking. The net effect on the RG flow is to give the gauge bosons $W^{\pm}$ and Z a mass.[5] Let us denote this mass scale collectively as M. For RG scales $\mu > M$, the $W^{\pm}$ and Z gauge bosons participate in the running of the gauge coupling as in the one-loop formulae above. However, for $\mu < M$, the $W^{\pm}$ and Z gauge bosons decouple from the effective theory and the appropriate gauge coupling which controls the weak interactions no longer runs and is frozen. This is why, unlike QCD, the weak coupling does not become large in the IR. Of course when the W and Z are decoupled they leave relics in the low-energy effective theory in the form of the weak four fermion interactions. The electro-magnetic coupling continues to run in the IR due to the presence of light charged particles. The running eventually stops at the mass of the electron, the lightest charged particle.

Now we consider the running of the three gauge couplings into the UV. The running is described by (4.36) where we take μ' to be an accessible energy scale around the mass of the Z, at which the values of the coupling can be fixed by using experimental data. Plotting the running into the UV yields[6]

---

[5] It is an important feature of the electro-weak sector that the Z boson is actually a mixture of the SU(2) neutral gauge boson and the U(1) gauge boson. The photon is then the appropriate orthogonal mixture.

[6] The data is taken from Amaldi et al. (1991).

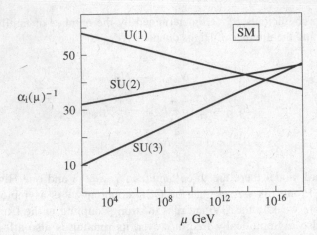

The graph shows that the 3 couplings converge at around $\mu \sim 10^{15}$ GeV. The matching is even more impressive in a supersymmetric extension of the standard model. In supersymmetric theories, described in Chap. 5, each particle has a super-partner which also contributes to the running. For instance, for the simplest supersymmetric extension of the standard model the running looks even more suggestive of a unification at high energies:

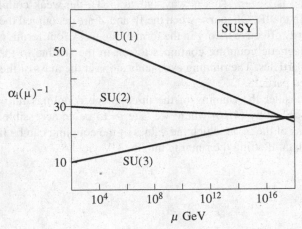

It is tempting to interpret this as evidence in favour of both supersymmetry and also of the idea of grand unification even though the simplest supersymmetric extension of the standard model has now been effectively ruled out by the latest data from the Large Hadron Collider. The idea of grand unification is that, since the 3 gauge couplings all converge on a single value, it is natural to suppose that at this scale the three gauge groups are actually unified into a single large gauge group G which contains $SU(3) \times SU(2) \times U(1)$ as a subgroup. This then requires that all the couplings at the unification scale to be equal. The standard model gauge group then arises via a symmetry breaking effect involving a new set of Higgs fields. Notice that in this scenario, the Landau pole of QED is repaired because beyond the unification scale, the QED coupling runs as the single coupling of the unified gauge theory G, which typically will be asymptotically free and perfectly well defined in the UV. This is an example of how an effective theory with a diverging backwards RG trajectory can be consistently "completed" in UV:

UV Completion

We have seen that an essential requirement in order to take a continuum limit in a QFT is that the theory lies on a renormalized trajectory, that is the RG flow out of some UV fixed point. For two theories that we have met, ϕ^4 scalar theory and QED, both in $d = 4$, there is no renormalized trajectory issuing from a UV fixed point and when we try to take the continuum limit there is a Landau pole which means that as we run the RG backwards the coupling becomes singular at some finite μ.

If this was the last word on continuum limits, then this would seem to be rather disastrous for the standard model which includes QED and the Higgs scalars. However, we have already seen that as the RG scale flows into the IR and passes the mass of a particle into the regime $\mu < m$, the particle is essentially redundant as far as the running is concerned and in RG schemes like $\overline{\text{MS}}$ is actually removed by hand from the theory (although it is important to remember that it leaves its mark in the form of irrelevant operators suppressed by powers of $1/m$ which can have an observable effect). So in the theories with Landau poles, it may be that there is a good *UV completion* which requires new particles and their fields as we flow into the UV. For example, for QED we have already seen how a consistent UV completion can be obtained by a grand unification scenario.

What these situations illustrate is that theories which have Landau poles are in a sense more interesting because they predict their own demise and force us to confront the issue of UV completion.

Bibliographical Notes

The running of the coupling in QED and the fact that it becomes large in the UV was first appreciated by Landau (1955). The running of gauge couplings in a non-abelian gauge theory was first described in the classic papers of Gross and Wilczek (1973) and Politzer (1973) and is fully described in the text-book by Peskin and Schroeder (1995). The data for the graphs that we used to show the running of the gauge couplings in the standard model and the implications for grand unification are taken from Amaldi et al. (1991). The subtle issues of decoupling in mass dependent and independent schemes are described in the excellent review article on effective field theories by Manohar (1996).

References

Amaldi, U., de Boer, W., Furstenau, H.: Comparison of grand unified theories with electroweak and strong coupling constants measured at LEP. Phys. Lett. B **260**, 447 (1991)

Banks, T., Zaks, A.: On the phase structure of vector-like gauge theories with massless fermions. Nucl. Phys. B **196**, 189 (1982)

Gross, D.J., Wilczek, F.: Ultraviolet behavior of non-abelian gauge theories. Phys. Rev. Lett. **30**, 1343 (1973)

Landau, L.D.: In: Pauli, W. (ed.) Niels Bohr and the Development of Physics, p. 52. Pergamon Press, London, (1955)

Manohar, A.V.: In: Schladming 1996, Perturbative and Nonperturbative Aspects of Quantum Field Theory, pp 311–362 [hep-ph/9606222]

Peskin, M.E., Schroeder, D.V.: An Introduction to Quantum Field Theory. Addison-Wesley, Reading (1995)

Politzer, H.D.: Reliable perturbative results for strong interactions. Phys. Rev. Lett. **30**, 1346 (1973)

Chapter 5
RG and Supersymmetry

QFTs with supersymmetry (SUSY) have some remarkable properties that make them particularly interesting to study from the point-of-view of the RG. For instance, one can prove the existence of many non-trivial fixed points of the RG in such theories in $d = 4$. Our discussion of SUSY theories will be geared towards describing some of these fixed points and the associated RG structure at the expense of many other aspects of SUSY theories. In particular, we shall use a description in terms of component fields rather than introduce the whole paraphernalia of superspace and super-fields.

In a SUSY theory, fields are collected into multiplets of supersymmetry which consist of fields with different spin. In $d = 4$, which we will stick to exclusively, there are 2 basic multiplets: (i) a chiral multiplet consisting of a complex scalar, a Weyl fermion and an auxiliary complex scalar field

$$\Phi = (\phi, \psi_\alpha, F), \tag{5.1}$$

(ii) a vector multiplet consisting of a gauge field, an adjoint-valued Weyl fermion and an adjoint-valued auxiliary real scalar field

$$V = (A_\mu, \lambda_\alpha, D). \tag{5.2}$$

All our conventions are taken from the text-book by Wess and Bagger (1992).

5.1 Theories of Chiral Multiplets: Wess-Zumino Models

Wess-Zumino models are constructed from chiral multiplets Φ_i with a basic SUSY invariant kinetic term of the form[1]

---

[1] The kinetic term can be generalized to all the terms with two derivatives:

$$\mathscr{L}_{\text{kin}} = -|\partial_\mu \phi_i|^2 + i\bar{\psi}_i \bar{\sigma}^\mu \partial_\mu \psi_i + |F_i|^2. \tag{5.4}$$

Note that the auxiliary fields F_i do not have kinetic terms and their only rôle in the theory is to simplify the SUSY structure. The interactions are determined by the super-potential $W(\phi_i)$, a function of the fields ϕ_i, but not their complex conjugates:

$$\mathscr{L}_{\text{int}} = F_i \frac{\partial W}{\partial \phi_i} - \frac{1}{2} \frac{\partial^2 W}{\partial \phi_i \partial \phi_j} \psi_i \psi_j + \text{c.c.} \tag{5.5}$$

(c.c.=complex conjugate.) Notice that since the auxiliary fields F_i only appear quadratically in the Lagrangian, they can trivially be "integrated out" exactly by using their equations-of-motion

$$F_i = -\frac{\partial W(\phi_j^*)}{\partial \phi_i^*} \tag{5.6}$$

and substituting this back into the action to leave a net potential on the scalar fields of the form

$$V(\phi_i, \phi_i^*) = \sum_i \left| \frac{\partial W}{\partial \phi_i} \right|^2. \tag{5.7}$$

The theory is invariant under the infinitesimal SUSY transformations involving a space-time constant Grassmann spinor parameter ξ_α:

$$\begin{aligned}
\delta \phi_i &= \xi \psi_i, \\
\delta \psi_i &= i\sigma^\mu \bar{\xi} \partial_\mu \phi_i + \xi F_i, \\
\delta F_i &= i\bar{\xi} \bar{\sigma}^\mu \partial_\mu \psi_i.
\end{aligned} \tag{5.8}$$

There is an important distinction between the kinetic and interaction terms: $\mathscr{L}_{\text{kin}}$ are "D terms", which are real functions of the fields and their complex conjugates, while $\mathscr{L}_{\text{int}}$ are "F terms", consisting of the sum of a part which is holomorphic in the fields and the couplings and the complex conjugate term which has all quantities replaced by the corresponding anti-holomorphic ones. In particular, the super-potential $W(\phi_i)$ is a holomorphic function of the ϕ_i and also the couplings g_n, i.e. $W(\phi_i)$ does not depend on ϕ_i^* or g_n^*. It is this underlying holomorphic structure of the F-terms that plays a pivotal rôle in leading to exact RG statements in SUSY theories.

(Footnote 1 continued)

$$\mathscr{L}_{\text{kin}} = -\frac{\partial^2 K}{\partial \phi_i \partial \phi_j^*} \partial_\mu \phi_i \partial^\mu \phi_j^* + \text{terms involving } (\psi, F), \tag{5.3}$$

determined by a real function $K(\phi, \phi^*)$, the *Kähler potential*.

For example, for a ϕ^4 type model the super-potential has the form

$$W(\phi) = \frac{1}{2}m\phi^2 + \frac{1}{3}\lambda\phi^3, \tag{5.9}$$

where the mass and couplings m and λ are in general complex. In this case, after integrating out the single auxiliary field F, the interaction terms are

$$\mathscr{L}_{\text{int}} = -\left|m\phi + \lambda\phi^2\right|^2 - \frac{m}{2}\psi\psi - \frac{m^*}{2}\bar{\psi}\bar{\psi} - \lambda\phi\psi\psi - \lambda\phi^*\bar{\psi}\bar{\psi}. \tag{5.10}$$

The first term here is the potential (5.7) while the final two terms are Yukawa interactions between the fermions and scalars.

SUSY Non-Renormalization

From the point-of-view of RG, the key fact is that couplings in the super-potential do not change when we change the RG scale μ. The proof of this is very simple. If we assume that the cut off is consistent with SUSY then when we integrate out modes of the fields, as we decrease μ, the structure of the super-potential must be preserved in the sense that it can only depend on the holomorphic couplings. In other words, the beta functions of the couplings in the super-potential must preserve the holomorphic structure so that if we denote the holomorphic and anti-holomorphic couplings as g_i and $\bar{g}_i$, then we have

$$\mu\frac{dg_i}{d\mu} = \beta_i(g_j), \qquad \mu\frac{d\bar{g}_i}{d\mu} = \beta_i(\bar{g}_j), \tag{5.11}$$

rather than $\beta_i(g_j, \bar{g}_j)$. Imagine for a moment an unphysical world in which the holomorphic couplings are not complex conjugates. In fact when calculating the beta functions of the holomorphic couplings we can imagine an extreme world in which the anti-holomorphic couplings actually vanish $\bar{g}_i = 0$. Since the propagators of the theory are D-terms they inevitably join a holomorphic leg in a Feynman diagram to an anti-holormophic leg. But since there are no couplings involving the anti-holomorphic fields there are no perturbative Feyman diagrams that one can write down that lead to a renormalization of the couplings g_i. It seems we can then infer that since $\beta_i(g_j)$ does not depend on $\bar{g}_j$ this must also be true in the physical theory where $\bar{g}_j = g_j^*$, the complex conjugate.

The non-renormalization theorem does *not* mean that the couplings in the super-potential do not have any quantum contributions to RG flow because there can be non-trivial wave-function renormalization. The latter involves the kinetic terms, which are D-terms and so do not have a holomorphic structure and the arguments above do not apply. In fact, for a term of the form $W \sim \mu^{3-n}\lambda\phi^n$, the RG equation becomes

$$W(Z(\mu)^{1/2}\phi; \mu, \lambda(\mu)) = W(Z(\mu')^{1/2}\phi; \mu', \lambda(\mu')) \tag{5.12}$$

and so under an RG flow

$$\mu^{3-n}Z(\mu)^{n/2}\lambda(\mu) = \mu'^{3-n}Z(\mu')^{n/2}\lambda(\mu'), \tag{5.13}$$

from which we extract the beta-function

$$\mu\frac{d\lambda}{d\mu} = (-3 + n + n\gamma)\lambda, \tag{5.14}$$

where γ is the common anomalous dimension of all fields in the chiral multiplet. SUSY ensures that all fields in a multiplet have the same wave-function renormalization. One implication of this result is that the scaling dimension of the *composite operator* ϕ^n is just equal to the sum of the scaling dimension of the individual operators ϕ:[2]

$$\Delta_{\phi^n} = n(1 + \gamma) = n\Delta_\phi. \tag{5.15}$$

On the contrary, SUSY does *not* constrain the running of D-term couplings.

The fact that the super-potential is only renormalized by wave-function renormalization has a very important application. The vacua of a theory are determined by minimizing the effective potential. This latter quantity is, as argued previously, the potential in the Wilsonian effective action in the limit $\mu \to 0$. In a SUSY theory, the potential is given by (5.7), and so the global minima of V correspond to critical points of the super-potential, the "F-flatness" condition:

$$F_i^* = -\frac{\partial W(\phi_j)}{\partial \phi_i} = 0. \tag{5.16}$$

Such vacua can be shown to preserve SUSY because $V = 0$, whereas if the minima have $V > 0$ then this is a sign that SUSY is spontaneously broken. Now we come to the important bit: as we decrease the cut off $\mu \to 0$, if we have a solution ϕ_i of (5.16) at the original cut off, then we still have a solution, differing only by wave-function renormalization, $\phi_i \to Z_i(\mu)^{1/2}\phi_i$, and so the SUSY preserving vacua of the theory can be determined from the super-potential alone.[3]

For example, the theory with a super-potential (5.9) has critical points at $\phi = 0$ and $\phi = -m/\lambda$, and by the argument above these minima will be vacua of the full QFT with the fields only undergoing wave-function renormalization once the quantum corrections are taken into account. It is also possible to construct models which do not have any SUSY vacua.

---

[2] It is a key aspect of the RG that the scaling dimension of a general composite operator $\mathcal{O} = \mathcal{O}_1 \cdots \mathcal{O}_p$ is usually *not* the sum $\Delta_\mathcal{O} \neq \Delta_{\mathcal{O}_1} + \cdots + \Delta_{\mathcal{O}_p}$. For instance, for ϕ^4 theory around the Wilson-Fisher fixed point one has $\Delta_{\phi^4} \neq 2\Delta_{\phi^2}$.

[3] There is a caveat to this in that it depends on the fact that kinetic terms are suitably well behaved at the minima.

Vacuum Moduli Spaces

In a non-SUSY QFT, vacua of the theory are generally discrete points because quantum corrections and RG flow of the effective potential generically lift any flat directions of the potential to leave isolated minima. In certain cases flat directions can arise if there is a broken symmetry but then the inequivalent vacua are essentially equivalent since they are related by symmetry transformations. In a SUSY theory, since the effective potential is not renormalized, degenerate spaces of vacua are not, in general, lifted. So in many cases, SUSY theories have non-trivial Vacuum Moduli Spaces $\mathfrak{M}$ whose points are labelled by the VEVs of the scalar fields in the theory and are not related by any symmetry. The space $\mathfrak{M}$ is in many cases not a manifold; rather, it is a series of manifolds, or "branches", joined along subspaces of lower dimension.

As an example consider a model with 3 chiral multiplets Φ_i, $i = 1, 2, 3$ with a super-potential $W = \lambda \phi_1 \phi_2 \phi_3$. The SUSY vacua are determined by the equations $\phi_1 \phi_2 = \phi_2 \phi_3 = \phi_3 \phi_1 = 0$ and so there are 3 branches: (i) $\phi_1 = \phi_2 = 0$, ϕ_3 arbitrary, (ii) $\phi_2 = \phi_3 = 0$, ϕ_1 arbitrary, and (iii) $\phi_3 = \phi_1 = 0$, ϕ_2 arbitrary. These three branches are joined at the point $\phi_1 = \phi_2 = \phi_3 = 0$. The structure of $\mathfrak{M}$ is schematically

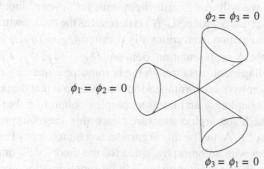

5.2 SUSY Gauge Theories

A SUSY gauge theory is constructed from a kinetic term which, unlike the chiral multiplet, is an F-term:

$$\mathscr{L}_{\text{kin}}(V) = \left(\frac{1}{2g^2} + \frac{\theta}{16\pi^2 i} \right) \mathscr{L}(V) + \text{c.c.}, \tag{5.17}$$

$$\mathscr{L}(V) = -\frac{1}{4} F_{\mu\nu}^a F^{a\mu\nu} + \frac{i}{8} \varepsilon_{\mu\nu\rho\sigma} F^{a\mu\nu} F^{a\rho\sigma} - i\bar{\lambda}^a \bar{\sigma}^\mu D_\mu \lambda^a + \frac{1}{2} D^a D^a.$$

Notice that the gauge coupling is naturally combined with the θ angle[4] to form a holomorphic coupling which is conveniently written as

$$\tau = \frac{4\pi i}{g^2} + \frac{\theta}{2\pi}. \tag{5.18}$$

If we couple a vector super-field to a chiral super-field transforming in some representation r of the gauge group with generators T_{ij}^a, then a SUSY invariant action involves replacing all derivatives by covariant derivatives and including the additional interactions

$$\mathscr{L}_{\text{int}} = \frac{i}{\sqrt{2}} T_{ij}^a \left(\phi_i^* \psi_j \lambda^a - \phi_j \bar{\lambda}^a \bar{\psi}_i \right) + D^a T_{ij}^a \phi_i^* \phi_j. \tag{5.19}$$

Notice that once the auxiliary fields D^a and F_i are integrated-out, the scalar fields now have a net potential

$$V(\phi_i, \phi_i^*) = \sum_i \left| \frac{\partial W}{\partial \phi_i} \right|^2 + \frac{1}{2} \sum_{ija} \left(\phi_i^* T_{ij}^a \phi_j \right)^2, \tag{5.20}$$

which generalizes (5.7). Henceforth, we shall think of the gauge index of ϕ as being implicit and treat it as a vector. If we have several matter fields transforming in representations r_f then we will distinguish them with a "flavour" index Φ_f. For instance, SUSY QCD with gauge group $SU(N)$ is defined as the theory with N_f chiral multiplets in the N representation, conventionally denoted $Q_f = (q_f, \psi_f, F_f)$, and N_f chiral multiplets in the $\bar{N}$ representation, denoted $\tilde{Q}_f = (\tilde{q}_f, \tilde{\psi}_f, \tilde{F}_f)$.

Since the gauge coupling manifests through the holomorphic quantity τ one wonders whether it has any non-trivial renormalization. The reason is that the perturbative expansion is in the real coupling g and not the complex coupling τ, but renormalization must preserve the holomorphic structure hence this seems to preclude any RG flow of τ and hence of g. Actually, this argument is a bit too hasty because one-loop running is consistent with holomorphy, since for the theory with only a vector multiplet, using (4.29) with one fermion in the adjoint representation, we have

$$\mu \frac{dg}{d\mu} = -\frac{3}{(4\pi)^2} C(G) g^3 \quad \implies \quad \mu \frac{d\tau}{d\mu} = \frac{3i}{2\pi} C(G). \tag{5.21}$$

This is consistent because θ does not run. So we conclude that the beta function of g is exact at the one-loop level in a SUSY gauge theory.

---

[4] The θ angle multiplies a term $\frac{1}{64\pi^2} \int d^4x \, \varepsilon_{\mu\nu\rho\sigma} F^{a\mu\nu} F^{a\rho\sigma}$ in the action. This integral, which computes the 2nd Chern Class of the gauge field, is topological, in the sense that for any smooth gauge configuration it is equal to $2\pi k$, for k an integer. Furthermore, this term does not contribute to the classical equations-of-motion. In the quantum theory which involves the Feynman sum over configurations, θ becomes physically meaningful and should be treated as another coupling in the theory.

Once we add chiral multiplets coupled to the vector multiplet, the beta function of g will get non-trivial contributions for two reasons. The first effect is simple: the one-loop coefficient receives additional contributions from the fields of the chiral multiplets. From (4.29), and taking into account the field content of the vector and chiral multiplets, we simply have to perform the replacement

$$C(G) \longrightarrow C(G) - \frac{1}{3} \sum_f C(r_f). \tag{5.22}$$

The second contribution is more subtle. Under the RG transformation, the kinetic term changes due to wave-function renormalization. If we start at some RG scale μ_0 with $Z = 1$, then as we lower the scale to μ

$$|\partial_\nu \phi|^2 \longrightarrow Z(\mu) |\partial_\nu \phi|^2. \tag{5.23}$$

At this point, we have to perform the re-scaling $\phi \to Z^{-1/2} \phi$ in order to return the kinetic term to its canonical form. When the re-scaling is performed, we should take into account the Jacobian arising from the measure of the functional integral:

$$J = Z^{-\mathcal{N}_{<\mu}/2}, \tag{5.24}$$

where $\mathcal{N}_{<\mu}$ is the number of modes with momentum less than the cut off μ. The number of modes can be calculated as in Chap. 2 by defining the Euclidean theory in a large box of space-time volume $\mathcal{V} = L^d$ and imposing periodic boundary conditions on the field. In this case the number of modes is the volume in Euclidean phase space divided by $(2\pi\hbar)^d$ (with $\hbar = 1$) giving

$$J = \exp\left[-\frac{1}{2} \log Z \, \mathcal{V} \int_{|p|<\mu} \frac{d^d p}{(2\pi)^d} \right] \tag{5.25}$$

$$= \exp\left[-\frac{1}{2} \log Z \, \mu^d \, \frac{\mathrm{Vol}(S^{d-1})}{(2\pi)^d} \mathcal{V} \right].$$

This is just a constant factor that we can safely ignore. However, when ϕ is coupled non-trivially to a gauge field, the Jacobian will depend on the background gauge field. The reason is that in order to preserve gauge invariance, the cut off must somehow be compatible with the covariant derivative $D_\nu = \partial_\nu + i A_\nu$ and this means that the whole cut-off procedure must involve the background gauge field in a non-trivial way. The proper way to do this is to put the cut off μ^2 on the eigenvalues of the covariant Laplace equation (in Euclidean space):

$$- D_\nu^2 \phi = \lambda \phi. \tag{5.26}$$

It is technically quite complicated to calculate the Jacobian this way; however, we can avoid this by making intelligent use of the chiral anomaly.[5]

The Chiral Anomaly

A chiral transformation on a spinor field takes

$$\psi \rightarrow e^{i\alpha}\psi, \quad \bar{\psi} \rightarrow e^{-i\alpha}\bar{\psi}. \tag{5.27}$$

The action is clearly invariant under this transformation, however, the regularized measure $\int [d\psi][d\bar{\psi}]$ is not. The point is that the cut-off procedure breaks the symmetry in the presence of a background gauge field. Under the transformation, the fermion part of the measure picks up a non-trivial Jacobian which yields the well-known chiral anomaly

$$[e^{i\alpha}d\psi][e^{-i\alpha}d\bar{\psi}] = [d\psi][d\bar{\psi}]\exp\left[-\frac{i\alpha C(r)}{32\pi^2}\int d^4x\, \varepsilon_{\mu\nu\rho\sigma}F^{a\mu\nu}F^{a\rho\sigma}\right]. \tag{5.28}$$

What is remarkable about the chiral anomaly is that it is an exact result to all orders in perturbation theory.

We can now use the chiral anomaly to calculate the re-scaling anomaly, or Jacobian, for a chiral multiplet by exploiting holomorphy and SUSY. If one compares the right-hand side of (5.28) with the SUSY kinetic term (5.17), whilst taking into account that the holomorphic and anti-holomorphic contributions to $\mathscr{L}_{\text{kin}}$ are separately invariant under SUSY, it is clear that (5.28) can be written in a way manifestly invariant under SUSY; namely,

$$
\begin{aligned}
J &= \exp\left\{\frac{\alpha C(r)}{8\pi^2}\int d^4x\left[\mathscr{L}(V) - \mathscr{L}^*(V)\right]\right\} \\
&= \exp\left\{-\frac{iC(r)}{8\pi^2}\int d^4x\left[(i\alpha\mathscr{L}(V)) + (i\alpha\mathscr{L}(V))^*\right]\right\}.
\end{aligned} \tag{5.29}
$$

Most of the terms here cancel to leave only the right-hand side of (5.28). However, we can now exploit the holographic split to generalize the chiral transformation to an arbitrary holomorphic re-scaling of the whole chiral multiplet:

$$\phi \rightarrow Z^{-1/2}\phi, \quad \psi \rightarrow Z^{-1/2}\psi, \quad F \rightarrow Z^{-1/2}F. \tag{5.30}$$

Up to an unimportant constant factor, the resulting Jacobian is then simply the right-hand side of (5.29) with $e^{i\alpha}$ replaced by $Z^{-1/2}$:

---

[5] The following discussion is taken from Arkani-Hamed and Murayama (1998).

$$J = \exp\left\{ \frac{iC(r)}{16\pi^2} \int d^4x \left[\left(\log Z \mathscr{L}(V) \right) + \left(\log \bar{Z} \mathscr{L}(V) \right)^* \right] \right\}. \tag{5.31}$$

The clever bit of the argument is that by analytic continuation this result must also be valid for a wave-function renormalization for which Z is real. Hence, we can calculate the effect of wave-function renormalization on the running of the gauge coupling. Under an infinitesimal RG transformation, the wave-function factor changes to $Z = 1 - 2\gamma \delta\mu/\mu$ and so the Jacobian is easily seen to be of the form

$$J = \exp\left\{ -\frac{iC(r)\gamma}{8\pi^2} \int d^4x \left[\mathscr{L}(V) + \mathscr{L}^*(V) \right] \frac{\delta\mu}{\mu} \right\}, \tag{5.32}$$

which corresponds to an additional contribution to the flow of the gauge coupling of

$$\delta\left(\frac{1}{g^2}\right) = \frac{C(r)\gamma}{4\pi^2} \cdot \frac{\delta\mu}{\mu},$$

$$\mu \frac{dg}{d\mu}\bigg|_{\text{additional}} = -\frac{g^3}{8\pi^2} C(r)\gamma. \tag{5.33}$$

Hence, the exact beta function for a model with a series of chiral multiplets in representations r_f of the gauge group is

$$\mu \frac{dg}{d\mu} = -\frac{g^3}{16\pi^2} \left(3C(G) - \sum_f C(r_f)(1 - 2\gamma_f) \right). \tag{5.34}$$

The above is an exact result, valid for the coupling that appears as $1/g^2$ in front of the gauge kinetic term. It is tempting to think that this is the same as the *canonical gauge coupling* g_c defined as the one that appears in the covariant derivatives $D_\mu = \partial_\mu + ig_c A_\mu$ because one can simply re-scale the gauge field $A_\mu \to g_c A_\mu$ (and all the other fields of the vector multiplet). But as we have found above, these kinds of re-scalings will lead to a non-trivial Jacobian coming from the measure of the functional integral. The up-shot of this is that we cannot simply identify the two couplings g and g_c. Unfortunately the trick that worked for a chiral multiplet does not work in a simple way here, since the chiral rotation of the gluino cannot be complexified to give the re-scaling anomaly of the vector multiplet. Consequently we just quote the result:[6]

$$J = \exp\left\{ \frac{iC(G)g^2}{16\pi^2} \log g_c \int d^4x \left(\mathscr{L}(V) + \mathscr{L}(V)^* \right) \frac{\delta\mu}{\mu} \right\}. \tag{5.35}$$

Notice that it has the same form as an adjoint-valued chiral multiplet but with the opposite sign and with $Z = g_c^2$. Now we make the re-scaling $A_\mu \to g_c A_\mu$, taking into account the additional factor above, and, thereby derive the following relation

---

[6] The calculation is explained in detail by Arkani-Hamed and Murayama (1998).

between the two definitions of the coupling:

$$\frac{g_c^2}{g^2} = 1 + \frac{C(G)}{4\pi^2} g_c^2 \log g_c, \tag{5.36}$$

which is the exact relation between the two definitions of the gauge coupling, g and g_c. The beta function of the canonical coupling then follows:

The NSVZ Beta Function

$$\mu\frac{dg_c}{d\mu} = \mu\frac{dg}{d\mu} \cdot \frac{g_c^3}{g^3} \cdot \frac{1}{1 - C(G)g_c^2/8\pi^2}$$

$$= -\frac{g_c^3}{16\pi^2} \cdot \frac{3C(G) - \sum_f C(r_f)(1 - 2\gamma_f)}{1 - C(G)g_c^2/8\pi^2}. \tag{5.37}$$

This is the famous Novikov-Shifman-Vainshtein-Zahakarov beta function for SUSY gauge theories (Novikov et al. 1983).

5.3 Vacuum Structure

Searching for the vacua of a SUSY gauge theory is more complicated than for a Wess-Zumino model. The potential term of a SUSY gauge theory (5.20) includes a contribution from the gauge sector which arises once the D field is integrated out. Clearly, as in the Wess-Zumino case, it is bounded below by 0 and in order to preserve SUSY we need $V = 0$ and so the conditions for a SUSY vacuum are

$$\frac{\partial W}{\partial \phi_f} = 0, \qquad \sum_f \phi_f^\dagger T^a \phi_f = 0. \tag{5.38}$$

These are known as the F- and D-term equations, respectively, and have to be solved modulo gauge transformations.

To see how this works, consider SUSY QED with chiral fields Q_1 and Q_2 of charge $+1$ and $\tilde{Q}_1$ and $\tilde{Q}_2$ of charge -1. Furthermore, let us suppose that the theory has no super-potential. Hence, only the D-term equation is non-trivial:

$$|q_1|^2 + |q_2|^2 - |\tilde{q}_1|^2 - |\tilde{q}_2|^2 = 0. \tag{5.39}$$

One way to solve this is to use gauge transformations to make q_1 real and then the D-term equation determines, say, q_1. This breaks down when $q_1 = 0$, but then we can use q_2 instead, etc. All-in-all, modulo gauge transformations, we have a 3-complex dimensional space of solutions which defines the vacuum moduli space $\mathfrak{M}$. At any point in $\mathfrak{M}$, besides the origin, the VEV of the scalar fields breaks the gauge group and the photon gains a mass by the Higgs mechanism.

It is an important theorem that any solution of the F- and D-term equations modulo gauge transformations is equivalent to a solution of the F-term equations *alone* modulo *complexified* gauge transformations.

Complexified Gauge Transformations

We present the proof in the abelian case $G = \mathrm{U}(1)$. Suppose we have a solution of the F-term equations q_f (with charges e_f). The key point is that since the super-potential only depends on the holomorphic fields and not the anti-holomorphic ones, it is actually invariant not just under gauge transformations $q_f \to e^{i e_f \theta} q_f$ but also complexified gauge transformations $q_f \to \eta^{e_f} q_f$, for arbitrary complex η. On the other hand, the D-term

$$\sum_f e_f |q_f|^2 \tag{5.40}$$

is not invariant under this complexified transformation. The strategy is then to use the complexified gauge transformation, with η real, to find a solution of the D-term equation. With $\check{q}_f = \eta^{e_f} q_f$, this is

$$\sum_f e_f |\check{q}_f|^2 = \sum_f e_f \eta^{2 e_f} |q_f|^2 = \frac{1}{2} \frac{\partial}{\partial \eta} \sum_f \eta^{2 e_f} |q_f|^2 = 0. \tag{5.41}$$

It is always possible to find an η which does this because for the function $\sum_f \eta^{2 e_f} |q_f|^2$ either (i) e_f all have the same sign, in which case $\eta \to 0$ or ∞ depending on the sign of the charges, or (ii) if some of the e_f have opposite sign $\sum_f \eta^{2 e_f} |q_f|^2$ goes to ∞ for $\eta \to 0$ or ∞, and hence has a minimum as a function of η for some finite $\eta > 0$. The generalization to the non-abelian case is reasonably straightforward: see (Wess and Bagger, 1992) p. 57.

This theorem is useful because it survives quantum corrections. In particular, although the D-terms are renormalized non-trivially, we can ignore them if we are only interested in the SUSY vacua of theory. In the present case, complexified gauge transformations act as

$$q_1 \to \lambda q_1, \quad q_2 \to \lambda q_2, \quad \tilde{q}_1 \to \lambda^{-1} \tilde{q}_1, \quad \tilde{q}_2 \to \lambda^{-1} \tilde{q}_2. \tag{5.42}$$

So for the example above, which has no F-term conditions, we can find $\mathfrak{M}$ by finding a set of coordinates which are gauge invariant in a complexified sense: in this case $z_1 = q_1\tilde{q}_1, z_2 = q_1\tilde{q}_2, z_3 = q_2\tilde{q}_1$ and $z_4 = q_2\tilde{q}_2$. However, these coordinates are not all independent, a short-coming that can be remedied by imposing the condition

$$z_1 z_4 = z_2 z_3. \tag{5.43}$$

So the vacuum moduli space is an example of a *complex variety* defined by the above condition in $\mathbf{C}^4$. In this case the complex variety is actually a *conifold* rather than a manifold since it is singular at $z_i = 0$ where the U(1) gauge symmetry is restored.

5.4 RG Fixed Points

It is a remarkable fact that one can prove the existence of many non-trivial RG fixed points of $d = 4$ SUSY theories. These fixed points have a SUSY extension of conformal invariance and are consequently super-conformal field theories. The existence of these fixed points, or more generally higher-dimensional fixed manifolds, rests on the special properties that we have established for the renormalization of the super-potential (5.14) and for the gauge coupling (5.34) or (5.37).

First of all, consider a Wess-Zumino model. In that case, it is a fact that we will not prove that all the anomalous dimensions $\gamma_i \geq 0$ (with equality for a free theory). Then in order to have an interacting theory, we need terms in the super-potential that are at least cubic in the fields, and so it follows immediately that there cannot be non-trivial fixed points of the RG:

$$\mu \frac{d\lambda}{d\mu} = (-3 + n + n\gamma)\lambda > 0, \tag{5.44}$$

for $n \geq 3$ and $\gamma > 0$.

Now we consider a gauge theory by taking a vector and a series of chiral multiplets in representations r_f of the gauge group. In that case, there is no positivity conditions on the anomalous dimensions of the chiral multiplets. In order to have a fixed point, for each coupling in the super-potential of the form $\lambda \phi_{f_1} \cdots \phi_{f_p}$, we need from (5.14)

$$\sum_{i=1}^{p} \gamma_{f_i} = 3 - p \tag{5.45}$$

and for the gauge coupling (5.34)

$$3C(G) - \sum_f C(r_f)(1 - 2\gamma_f) = 0. \tag{5.46}$$

If there are n such cubic couplings λ_i in the super-potential then it appears that there are $n + 1$ equations for $n + 1$ unknowns $\{\lambda_i, g\}$. Generically, therefore, if solutions exist they will be discrete and unlikely to be within the reach of perturbation theory. Up till now we appear to be in exactly the same situation as in a generic non-SUSY QFT. However, in the SUSY case note that both conditions above involve the anomalous dimensions of the chiral multiplets and so there are special situations when the set of $n + 1$ equations are linearly dependent. In such a scenario there are non-trivial spaces of fixed points which can extend into the perturbative regime and therefore be rigorously established.

For example, suppose there are 3 chiral multiplets in the adjoint representation with a super-potential cubic in the fields which has sufficient symmetry to infer that all the anomalous dimensions are equal, $\gamma_f \equiv \gamma$. In this case, (5.45) and (5.46) are satisfied if the anomalous dimensions vanishes:

$$\gamma(\lambda_i, g) = 0 \tag{5.47}$$

which is a single condition implying the existence of a manifold of fixed points. Theories such as this are very special because they are actually *finite*.

Finite Theories

A theory is *finite* if there are no UV divergences in perturbation theory. In a SUSY theory, this means that the anomalous dimensions of all the chiral operators vanish and the beta function of the gauge coupling vanishes. The conditions are:

(i) $\gamma_f = 0$.
(ii) The super-potential must be cubic in the fields in order that the RG flow of the couplings in the super-potential vanish: see (5.14).
(iii) $3C(G) = \sum_f C(r_f)$ in order that the RG flow of the gauge coupling vanishes.

Notice that not every finite theory is a CFT since conformal invariance could be broken by VEVs for scalar fields if there is a moduli space of vacua. However, in such cases conformal invariance would be recovered in the UV. Neither is every conformal field theory a finite theory since the condition to be at a fixed point does not require the anomalous dimensions of fields to vanish.

For example, if the gauge group is $G = \mathrm{SU}(N)$ then there are three gauge invariant couplings one can write in the super-potential for three adjoint-valued fields ($N \times N$ traceless Hermitian matrices) which have enough symmetry to imply that all the anomalous dimensions are equal:

$$W = \mathrm{Tr}\left(\lambda_1 \phi_1 \phi_2 \phi_3 + \lambda_2 \phi_1 \phi_3 \phi_2 + \frac{\lambda_3}{3} \sum_{f=1}^{3} \phi_f^3\right). \tag{5.48}$$

Of course we need to check that the condition $\gamma(\lambda_1, \lambda_2, \lambda_3, g) = 0$ actually has solutions. The key to proving this is to establish that solutions exist in perturbation theory and then by continuity infer that the solutions also exist at strong coupling. To one-loop order it can be shown that the anomalous dimension is[7]

$$\gamma = \frac{C(G)}{64\pi^2} \left(\sum_f |\lambda_f|^2 - 4g^2 \right).$$
(5.49)

Although we will not give the proof of this it is easy to see the diagrams that contribute. For example, for the anomalous dimension of the scalar field, the following 1-loop diagrams contribute (plus diagrams involving the ghosts which we do not show):

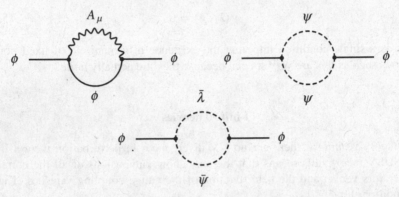

So γ receives positive contributions from the chiral multiplets and negative contributions from the vector multiplet. To simplify the discussion suppose that all the couplings are equal: $\lambda_f \sim \lambda$. In that case, there is a line of RG-fixed points at weak coupling when $\lambda = 2g/\sqrt{3}$. RG flow in the (λ, g) subspace is schematically of the form

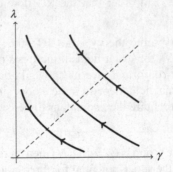

where the dotted line is a line of fixed points. The couplings away from the fixed line in the (λ, g) subspace are therefore irrelevant. One would expect that the fixed line extends by continuity into the region of strong coupling as well.

---

[7] The coupling here and in the following is generally the canonical gauge coupling, however, we will not distinguish between g and g_c from now on.

Once we take all the couplings into account, there is actually a 6-dimensional space of SCFTs (this comes from the complex λ_f, $f = 1, 2, 3$ and the gauge coupling subject to one condition).

Another class of finite theories is obtained by taking one chiral field Φ in the adjoint representation, $2N$ chiral multiplets Q_f in the N-dimensional defining representation of $SU(N)$ and $2N$ chiral multiplets $\widetilde{Q}_f$ in the conjugate representation, along with a super-potential of the form

$$W = \lambda \sum_{f=1}^{N} \widetilde{Q}_f \Phi Q_f. \tag{5.50}$$

The obvious symmetries of the theory are enough to ensure that all the chiral multiplets Q_f and $\widetilde{Q}_f$ have the same anomalous dimension. Hence, the beta functions are

$$\beta_g = -\frac{g^3 N}{32\pi}(\gamma_\Phi + 2\gamma_Q), \qquad \beta_\lambda = \gamma_\Phi + 2\gamma_Q, \tag{5.51}$$

which shows that there is a line of fixed point theories with $\gamma_\Phi(\lambda, g) + 2\gamma_Q(\lambda, g) = 0$. In this case, the super-CFTs are not necessarily finite since the anomalous dimensions do not need to separately vanish.

5.5 The Maximally SUSY Gauge Theory

A special class of these finite theories corresponds to taking a SUSY gauge theory with 3 chiral multiplets in the adjoint representation with a super-potential of the form (5.48) with $\lambda_1 = -\lambda_2 = \sqrt{2}g$ and $\lambda_3 = 0$, i.e.

$$W = \sqrt{2}g \text{Tr} \, \phi_1[\phi_2, \phi_3]. \tag{5.52}$$

Such theories have *extended* $\mathcal{N} = 4$ SUSY, the maximal amount of SUSY in $d = 4$ for theories that do not contain gravity. For later use, the potential of the theory has the form

$$V(\phi_i, \phi_i^*) = g^2 \sum_{ij=1}^{3} \text{Tr}(\phi_i \phi_j \phi_i^* \phi_j^* - \phi_i \phi_j \phi_j^* \phi_i^*). \tag{5.53}$$

The theory has a large global $SU(4)$ symmetry under which the 3 fermions from the chiral multiplets and the gluino together, transform as the 4-dimensional representation, while the 3 complex scalars ϕ_i can be written as 6 real scalars that transform in the antisymmetric 6-dimensional representation (or the vector of $SO(6) \simeq SU(4)$). These kinds of global symmetries of SUSY theories for which the fermions and scalars transform differently are known as R-symmetries. The $\mathcal{N} = 4$ theories (with vanishing VEVs for the scalars) lie on a line of RG fixed points labelled by

g and are, therefore, conformally invariant. In four-dimensional space-time the conformally group is the non-compact orthogonal group $SO(2, 4)$. It is interesting to note that the symmetry groups $SO(2, 4)$ and $SO(6)$ are the isometry groups of five-dimensional anti-de Sitter space and a five-sphere, respectively. These facts are a clue pointing to the remarkable duality first proposed by Maldacena (1997) that the $\mathcal{N} = 4$ gauge theory with gauge group $SU(N)$ is equivalent to Type IIB string theory in ten-dimensional space-time on an $AdS_5 \times S^5$ space-time geometry. The fact that what appears to be a humble four-dimensional gauge theory can encode all the rich dynamics of a ten-dimensional gravitational string theory must stand as one of the most far-reaching results in theoretical physics. Since the gauge theory lies at a fixed point of the RG and is, therefore, a CFT, along with the fact that the space-time geometry of the dual theory involves an anti de-Sitter space-time, explains why the relation between the two is known as the AdS/CFT correspondence. The rôle of the RG in the AdS/CFT correspondence is itself an interesting and active area for research; however, here we will limit ourselves to describing how the anomalous dimensions of certain operators of the CFT gauge theory are related to the masses of states in the dual string theory.

The gauge theory has two parameters, g and N, the gauge coupling and the specifier of the gauge group $SU(N)$. These are mapped to two parameters of the dual string theory. The first is the *string coupling* $g_s = g^2/4\pi$ which organizes string perturbation theory and the second is the radius of the $AdS_5 \times S^5$ geometry in string units $\sqrt{g^2N}$. This relation motivates the 't Hooft limit of the gauge theory which involves $N \to \infty$ with $\lambda = g^2N$ fixed. This is the so-called *planar* limit on the gauge theory side since only planar Feynman diagrams, those which can be drawn on the plane with no over-crossings, survive in the perturbation expansion, while on the string side $g_s \to 0$ and so it corresponds to non-interacting strings moving in an $AdS_5 \times S^5$ geometry. Perturbation theory on the gauge theory side requires λ to be small, which is the limit on the string side for which the geometry is highly curved. On the contrary, strong coupling, that is large λ, in the gauge theory corresponds to the regime where strings in the dual move on a weakly curved space-time geometry. Loosely speaking, a tractable regime of the gauge theory side is mapped to a difficult regime of the string theory and vice-versa.

Part of the "dictionary" between the two sides of the correspondence is that single trace composite operators in the gauge theory of the form

$$\mathcal{O}(x) = \text{Tr}(a_1 a_2 \cdots a_L), \tag{5.54}$$

where the a_i are one of the fundamental fields, correspond to states of a single string in the dual. Moreover the scaling dimension of the operator in the gauge theory equals the mass of the associated string state:

$$\mathcal{M}_{\text{string}} = \Delta_{\mathcal{O}}. \tag{5.55}$$

So perturbative calculations of the anomalous dimensions of operators in the gauge theory tells us directly about the spectrum of strings moving in a highly curved $AdS \times S^5$ geometry.

There have been some remarkable developments in calculating these anomalous dimensions in the gauge theory and matching them to the energies of string states. Here, we will consider the problem of calculating the anomalous dimensions of single trace operators made up of just the two basic fields ϕ_1 and ϕ_2 to one-loop in planar perturbation theory (perturbation theory in λ with $N = \infty$, a limit that suppresses non-planar contributions):

$$\mathcal{O} = \text{Tr}\left(\phi_1\phi_1\phi_2\cdots\right). \tag{5.56}$$

If the operator has length L then the classical dimension is simply $d_{\mathcal{O}} = L$, so we expect

$$\Delta_{\mathcal{O}} = L + \lambda\Delta_1 + \lambda^2\Delta_2 + \cdots. \tag{5.57}$$

The problem is that the operators $\mathcal{O}_i$ with a given number J_1 of ϕ_1 and J_2 of ϕ_2, with $J_1 + J_2 = L$, are all expected to mix under the RG and so to find the anomalous dimensions we have the problem of diagonalizing a large matrix.

One way to calculate the anomalous dimensions of the class of such operators $\{\mathcal{O}_p\}$, with fixed J_1 and J_2, is to add them to the action with their own coupling constant:

$$S \longrightarrow S + \int d^d x \sum_p \mu^{4-L} g_p \mathcal{O}_p(x) \tag{5.58}$$

and then look at the flow of the couplings g_p in the effective potential to linear order in g_p. We follow exactly the same background field method that we used earlier and treat the operator terms as new vertices in the action with couplings g_p. The flow of the couplings g_p can be deduced by writing down Feynman diagrams with J_1 external ϕ_1 lines and J_2 external ϕ_2 lines with fluctuating fields on internal lines.

Since the anomalous dimension follows from

$$\mu\frac{dg_p}{d\mu} = (L - 4)g_p + \gamma_{pq}g_q + \cdots, \tag{5.59}$$

we only need look at diagrams which use the vertices g_p once. Consider the coupling associated to the operator

$$\text{Tr}\left(\phi_{i_1}\phi_{i_2}\cdots\phi_{i_L}\right), \tag{5.60}$$

where each i_ℓ is either 1 or 2. At one-loop level, we have the diagrams

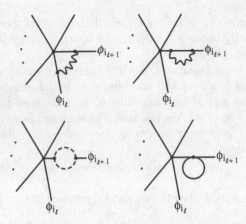

The first diagram here has to be summed over all neighbouring pairs (but not non-neighbouring pairs, since these would be non-planar diagrams suppressed by powers of $1/N$), while the others are to be summed over all L legs. The third involves a fermion loop and the fourth a scalar loop. The important point about these diagrams is that they do not change the "flavour" of the legs of the vertex. In other words, whatever their contribution to the anomalous dimension matrix is proportional to the identity in the space of (J_1, J_2) operators. We shall shortly argue that these contributions, which we write $C_1 \mathbf{1}$, actually vanish, so $C_1 = 0$.

The remaining diagrams involve using the quartic coupling in the scalar potential to tie two adjacent legs together. The potential 5.53 contains the terms

$$V = 2g^2 \mathrm{Tr}\big(\phi_1\phi_2\phi_1^*\phi_2^* - \phi_1\phi_2\phi_2^*\phi_1^* - \phi_2\phi_1\phi_1^*\phi_2^* + \phi_2\phi_1\phi_2^*\phi_1^*\big) + \cdots , \quad (5.61)$$

and so we see immediately that these interactions can be used to form two additional one-loop diagrams when two adjacent legs are different, either $\phi_1\phi_2$, as shown, or $\phi_2\phi_1$:

The absolute contribution is simple to calculate, but even without explicit calculation it is easy to see from the potential that these contributions come with a relative -1.

Putting all this together, we have found that the two sets of one-loop contributions to the anomalous dimension matrix, or "operator", can be written neatly as

$$\gamma = \sum_{\ell=1}^{L} \Big\{ C_1 \mathbf{1}_\ell + C_2 \big(\mathbf{1}_\ell - \mathbf{P}_\ell\big) \Big\}. \quad (5.62)$$

where $\mathbf{P}_\ell$ permutes the ℓth and $\ell + 1$th fields in the operator and $\mathbf{1}_\ell$ is the identity:

$$\mathbf{P}_\ell \operatorname{Tr}\left(\cdots \phi_{i_\ell} \phi_{i_{\ell+1}} \cdots\right) = \operatorname{Tr}\left(\cdots \phi_{i_{\ell+1}} \phi_{i_\ell} \cdots\right). \tag{5.63}$$

We also identify labels $\ell + L \equiv \ell$ due to the cyclicity of the trace.

Now we can pin down C_1 without having to calculate all the loop diagrams explicitly by using the fact that the operator $\operatorname{Tr} \phi_1^L$ is of a special kind known as a Bogomol'nyi-Prasad-Sommerfeld (BPS) operator which is known to be protected against all quantum corrections, this means that $\Delta = L$ to all orders in the perturbative expansion and so the anomalous dimension must vanish. Since from (5.62) we have $\gamma \operatorname{Tr} \phi_1^L = L C_1 \operatorname{Tr} \phi_1^L$, it must be that C_1 vanishes. This turns out to be verified when one actually calculates the one-loop graphs directly and, in addition, one finds that

$$C_2 = \frac{\lambda}{4\pi^2}. \tag{5.64}$$

Interestingly the resulting anomalous operator γ, up to some overall scaling, is identical to the Hamiltonian of the so-called XXX spin chain, a quantum mechanical model of L spins by identifying $|\uparrow\rangle \equiv \phi_1$ and $|\downarrow\rangle \equiv \phi_2$. Each operator of fixed length corresponds to a state of the spin chain:

$$\operatorname{Tr}\left(\phi_1 \phi_1 \phi_2 \phi_1 \phi_2 \phi_2 \phi_1\right) \longleftrightarrow |\uparrow\uparrow\downarrow\uparrow\downarrow\downarrow\uparrow\rangle. \tag{5.65}$$

The problem of finding the anomalous dimensions and hence the spectrum of string states then is identical to the problem of finding the eigenstates of the spin chain, a problem that was solved by Bethe (1931) by means of what we now call the *Bethe Ansatz* which reflects the fact that the problem is in the special class of integrable systems. This observation is just the beginning of the fascinating story of the underlying integrability of the $\mathcal{N} = 4$ gauge theory and the AdS/CFT correspondence.

Bibliographic Notes

A standard reference for the construction of SUSY theories from the viewpoint of superspace is the book of Wess and Bagger (1992). A distinctive, but excellent, treatment of SUSY theories is given by Strassler (2003). Our treatment of the RG flow of the gauge coupling and in particular the distinction between the two definitions of the gauge coupling is taken directly from the clear exposition by Arkani-Hamed and Murayama (1997, 1998). The exact beta function of SUSY gauge theories first appeared in the work of Novikov et al. (1983). The non-trivial fixed RG points that we described follows the pioneering work of Leigh and Strassler (1995). The story of the integrability that underlies the AdS/CFT correspondence is summarized in the series of review articles (Beisert et al. 2010).

References

Aharony, O., Gubser, S.S., Maldacena, J.M., Ooguri, H., Oz, Y.: Large N field theories, string theory and gravity. Phys. Rept. **323**, 183 (2000). [hep-th/9905111]

Arkani-Hamed, N., Murayama, H.: Holomorphy, rescaling anomalies and exact beta functions in supersymmetric gauge theories. JHEP **0006**, 030 (2000). [hep-th/9707133]

Arkani-Hamed, N., Murayama, H.: Renormalization group invariance of exact results in supersymmetric gauge theories. Phys. Rev. D **57**, 6638 (1998). [hep-th/9705189]

Beisert, N., Ahn, C., Alday, L.F., Bajnok, Z., Drummond, J.M., Freyhult, L., Gromov, N., Janik, R.A., Kazakov, V., Klose, T., Korchemsky, G.P., Kristjansen, C., Magro, M., McLoughlin, T., Minahan, J.A., Nepomechie, R.I., Rej, A., Roiban, R., Schafer-Nameki, S., Sieg, C., Staudacher, M., Torrielli, A., Tseytlin, A.A., Vieira, P., Volin, D., Zoubos, K.: Review of AdS/CFT Integrability: An Overview. arXiv:1012.3982 (2010). [hep-th]

Bethe, H.: Zur Theorie der Metalle. I. Eigenwerte und Eigenfunktionen der linearen Atomkette. Zeitschrift fr Physik **A71**, 205 (1931)

Leigh, R.G., Strassler, M.J.: Exactly marginal operators and duality infour-dimensional N=1 supersymmetric gauge theory. Nucl. Phys. B **447**, 95 (1995). [hep-th/9503121]

Maldacena, J.M.: The Large N limit of superconformal field theories and supergravity. Adv. Theor. Math. Phys. **2**, 231 (1998). [Int. J. Theor. Phys. **38**, 1113 (1999)][hep-th/9711200]

Novikov, V.A., Shifman, M.A., Vainshtein, A.I., Zakharov, V.I.: Exact Gell-Mann-Low function of supersymmetric Yang-Mills theories from instanton calculus. Nucl. Phys. B **229**, 381 (1983)

Strassler, M.J.: An unorthodox introduction to supersymmetric gauge theory. hep-th/0309149 (2003)

Wess, J., Bagger, J.: Supersymmetry and Supergravity. p. 259. Princeton University, USA (1992)